Limit Theorems For Sums Of Exchangeable Random Variables

Other Volumes in the
Rowman & Allanheld Probability and Statistics Series
Paul Switzer and Morris H. DeGroot, General Editors

Biostatistical Microcomputing in Pascal
Jerome H. Klotz and R. Daniel Meyer

Introduction to Statistical Methods, Volume I
Basic Concepts and Models
Jagdish S. Rustagi

Introduction to Statistical Methods, Volume II
Applications to Life Sciences
Jagdish S. Rustagi

forthcoming

Bayesian Analysis and Uncertainty in Economic Theory
Richard M. Cyert and Morris H. DeGroot

Limit Theorems For Sums Of Exchangeable Random Variables

Robert L. Taylor
University of Georgia

Peter Z. Daffer
Hughes Aircraft Company

Ronald F. Patterson
Georgia State University

Rowman & Allanheld
PUBLISHERS

ROWMAN & ALLANHELD

Published in the United States of America in 1985
by Rowman & Allanheld, Publishers
(a division of Littlefield, Adams & Company)
81 Adams Drive, Totowa, New Jersey 07512

Library of Congress Cataloging-in-Publication Data

Taylor, Robert Lee, 1943—
Limit theorems for sums of exchangeable random
variables.

(Rowman & Allanheld probability and statistics series)
Bibliography
Includes index.
1. Random variables. 2. Limit theorems (Probability
theory) I.Daffer, Peter Z. II. Patterson, Ronald F.
III. Title. IV. Title: Exchangeable random variables.
V. Series.
QA273.T395 1985 519.2 85-18310
ISBN 0-8476-7435-5

85 86 87 / 10 9 8 7 6 5 4 3 2 1
Printed in the United States of America

TABLE OF CONTENTS

PREFACE

Recent interest in exchangeable random variables as an alternative to the random sample being independent, identically distributed random variables has inspired the study of asymptotic properties of exchangeable random variables. These notes attempt to provide a unified presentation of the limit theorems for exchangeable random variables which have been obtained to date, including some recent work of the authors and colleagues.

These notes are somewhat self-contained with a background knowledge of graduate level probability, statistics and real analysis being assumed. Statisticians and quantitative scientists using exchangeable random variables will find this material very useful. In addition, this text will be useful in introducing advanced graduate students and researchers to the area of exchangeability and related limit theorems.

The authors are grateful to many research colleagues whose materials are included in the bibliography and in part are presented in the text. In particular, thanks go to Persi Diaconis, Jim Lynch and David Aldous for suggestions and help relating to this manuscript and other research. The authors are also grateful to Hiroshi Inoue, Bill McCormick, and a reviewer for the careful reading and comments of a preliminary draft of this manuscript. Special thanks go to Mrs. Jean Power and Mrs. Molly A. Rema for the careful typing and general appearance of this manuscript.

GENERAL INTRODUCTION

Many statistical estimators are expressible in the form

$$S_n = \Sigma_{k=1}^{n} a_{nk} X_{nk}. \qquad (0.1)$$

where $\{a_{nk}\}$ are constants and $\{X_{nk}\}$ are random variables. Thus, the convergence to a constant in probability or with probability one for S_n is of interest in determining consistency of the estimators. In addition, convergence in law for S_n is essential in establishing asymptotic properties of these estimators. While it is not always possible to assume that the random variables $\{X_{nk} : k \geq 1\}$ are independent for each n, the sequence $\{X_{n1}, X_{n2}, \ldots\}$ can usually be assumed to be permutation invariant with respect to distributions and that the dependence among the variables decreases as the sample size, n, increases. This concept of permutation invariance with respect to distributions is formally defined as <u>exchangeable</u>. The mean, $\bar{Y}_n$, of a random sample $\{Y_1, \ldots, Y_n\}$ can easily be seen to be of the form (0.1). Letting $a_{nk} = \frac{1}{n-1}$ and $X_{nk} = (Y_k - \bar{Y}_n)^2$, for $1 \leq k \leq n$, shows that the sample variance is also of the form (0.1). Note that the random variables $\{X_{n1}, \ldots, X_{nn}\}$ are not independent but are symmetric in their dependence for each n. Taylor (1982 and 1983) showed that the kernel density estimators were of the form (0.1) when the random variables took values in a function space and obtained complete convergence consistency to the unknown probability density function when the bandwidths were constants (and hence each row of the array could be considered as independent, identically distributed random variables in a function space). Other authors have also effectively used function space techniques for general function estimation problems. However, when the bandwidths are allowed to vary with data, then the weights $\{a_{nk}\}$ become random variables in each row of the array $\{X_{nk}\}$ are no longer independent. In many estimation problems, the linear form in (0.1) involves possibly random weights and possibly dependent random variables which are generated from independent, identically distributed random variables.

Since many statistical problems are addressed in considering linear forms of exchangeable random variables, limit theorems for S_n are an important consideration. Blum, Chernoff, Rosenblatt and Teicher (1958), Buhlmann (1959) & (1960), Eagleson and Weber (1978), Weber (1980) and others have obtained laws of large numbers and central limit theorems for real-valued exchangeable random variables. Beck (1976) was successful in obtaining some Banach space results for mutually symmetric and unconditionally independent random variables. Taylor (to appear) obtained laws of large numbers for arrays of row-wise exchangeable random variables and showed that these results could be related to a measure of nonorthogonality. Daffer (1984) was successful in obtaining central limit theorems for weighted sums of exchangeable random variables in Banach spaces by using the martingale techniques of Weber (1980).

The purpose of this monograph is to provide a unified presentation of the limit theorems for exchangeable random variables. While an exhaustive presentation with detailed proofs is not possible in this monograph, statisticians, mathematicians, and other quantitative scientists should find the coverage more than adequate in understanding and using these concepts and results.

In an attempt to make the notes reasonably self-contained, Chapter I will consist of basic definitions, properties and results for exchangeable random variables and will serve as a reference for the results of Chapters II and III. Chapter IV serves a similar role with respect to Chapters V and VI for random elements (random variables with values in a function space). Examples are provided throughout the chapters, and problems are listed at the end of chapters to further illustrate some of the concepts.

In Chapter II central limit theorems are obtained using the de Finetti integral representation and martingale techniques. The convergent limit laws are weighted normals, and conditions are given for these limit laws to be centered normals. Both sequential and array type structures are contained in these results. In the array setting exchangeability is assumed row-wise. In this setting the martingale techniques have the advantage of applicability for finite exchangeability in each row whereas the de Finetti representation requires infinite exchangeability.

Chapter II contains results for the convergence of S_n in probability and with probability one. A key consideration in these results is a measure of nonorthogonality defined similar to the correlation coefficient. For an array of random variables $\{X_{nk}\}$ which are exchangeable in each row, the bivariate distributions in each row must be identical, and the correlation coefficient $\rho_n \equiv$ correlation coefficient of X_{n1} and X_{n2} (when it exists) plays a crucial role in determining when S_n converges stochastically. In an over-simplification with respect to other appropriate hypotheses, S_n converges in probability if $\rho_n \to 0$ and converges with probability one with additional requirements on the rate of convergence to 0. Again martingale techniques provide powerful alternatives to considering the behavior of ρ_n.

Many of the properties and results for random variables remian valid when the random variables can assume values in a function space. Unfortunately, there are some notable exceptions which have spawned rich and extensive research areas. In Chapter IV the Banach space setting for the ranges of the random variables is described. Basic definitions and properties are given with particular attention given to measurability and expectation properties and to geometric considerations of the Banach spaces which are required for the limit theorems.

Central limit theorems for exchangeable random elements in Banach spaces are given in Chapter V. Both approaches of Chapter II, the use of the de Finetti representation and the martingale techniques, carry over with appropriate modifications. The limiting distribution for these results is the Gaussian process, and the Banach spaces must satisfy a smoothness condition (with respect to the unit balls) of type 2 or 2-uniformly smoothable. The required smoothness conditions relate to and circumvent the notable exceptions described in Chapter IV.

When considering the convergence of S_n in probability or with probability one in a Banach space, the geometric considerations are not as restrictive. Several of these results are provided in Chapter VI. Concentration of the probabilities on subsets of the conditional means with respect to the de Finetti representation is a main consideration in several of these results. When the Banach space is the real numbers and appropriate moment conditions hold, this requirement agrees with requirement of ρ_n going to zero in Chapter III. Other techniques used in obtaining the results in Chapter VI again include similar martingale techniques for Banach spaces.

CHAPTER I

PRELIMINARY AND BACKGROUND MATERIAL

1.0 INTRODUCTION

This chapter introduces the concept of exchangeability and considers some of the problems when considering limit theorems for exchangeable random variables. Properties which are needed for developing central limit theorems in Chapters II and IV and laws of large numbers in Chapters III and VI are provided in this chapter. Some background in probability and statistics at a graduate level is assumed. For example, it is assumed that the reader is familiar with the distributional properties of random variables and the traditional central limit theorems and laws of large numbers for independent random variables.

In Section 1.1 it is shown that all joint distributions of exchangeable random variables are the same and hence exchangeable random variables are stationary. Example 1.1 provides a generic example of exchangeable random variables which illustrates problems with traditional limit theorems. Several other examples are given, and it is shown that exchangeability is preserved under Borel mappings.

Exchangeability is related to conditional independence given the permutable σ-field or the tail σ-field in Section 1.2. Property 1.2.4 shows that the average of exchangeable random variables is a conditional expectation with respect to a sequence of decreasing σ-fields. Finally, the limit theorems are related to the common correlation coefficient (provided it exists) of exchangeable random variables.

In Section 1.3 de Finetti's theorem is discussed, and the distinction between infinite exchangeability and finite exchangeability is carefully noted. For an infinite sequence of exchangeable random variables, de Finetti's theorem implies that the exchangeable random variables are mixtures of independent, identically distributed sequences of random variables. An example is given where the traditional strong law of large numbers holds for a sequence of exchangeable random variables where first absolute moments do not exist. Finally, several measurability questions relating to subsets of probabilities determined by conditional means are addressed, and the chapter is concluded by a section of problems which further illustrate the concepts.

1.1 DEFINITIONS AND BASIC PROPERTIES OF EXCHANGEABILITY

The following definitions, notations, and basic properties will be used throughout these notes. First, R will denote the real numbers, and $\mathcal{B}(R)$ will denote the Borel subsets or the σ-field generated by the intervals $(-\infty,a]$, $a \in R$, (or equivalently, generated by the open subsets of R). Random variables will usually be denoted by the capital letters X, Y, Z, etc., and (Ω,A,P) will denote a generic probability space throughout the notes. Distribution functions are denoted by F or F_X, and EX denotes the expected value of the random variable X.

Definition 1.1.1 A finite sequence of random variables $\{X_1, \ldots, X_n\}$ is said to be exchangeable if the joint distribution of $(X_1, \ldots, X_n)$ is invariant with respect to permutations of the indices 1, ..., n. A collection of random variables $\{X_\alpha : \alpha \in \Gamma\}$ is said to be exchangeable if every finite subset of $\{X_\alpha : \alpha \in \Gamma\}$ is exchangeable.

To illustrate the concept of exchangeability (sometimes referred to as interchangeability) and to relate it to the usual probability concepts, let $\{X_1,X_2\}$ be exchangeable random variables. The permutation invariance of the joint distribution yields $F_{X_1,X_2}(a,b) = F_{X_2,X_1}(a,b)$ for all $a,b \in R$. Taking the limit as $b \to \infty$ gives

$$F_{X_1}(a) = \lim_{b\to\infty} F_{X_1,X_2}(a,b) = \lim_{b\to\infty} F_{X_2,X_1}(a,b) = F_{X_2}(a).$$

Thus, exchangeable random variables are identically distributed. Moreover, all joint distributions must be the same for exchangeable random variables as the following proposition shows.

Proposition 1.1.1 Let $\{X_1, \ldots, X_n\}$ be exchangeable random variables. Then for any integer k, $1 \le k \le n$, and subset $\{i_1, \ldots, i_k\}$ of k distinct elements from $\{1, \ldots, n\}$

$$P[X_1 \varepsilon B_1, \ldots, X_k \varepsilon B_k] = P[X_{i_1} \varepsilon B_1, \ldots, X_{i_k} \varepsilon B_K] \quad (1.1.1)$$

for all $B_1, \ldots, B_k \varepsilon \mathcal{B}(R)$.

Proof Let π be a permutation on $\{1,2, \ldots, n\}$ such that $\pi_j = i_j$ for $1 \le j \le k$. Given $B_1, \ldots, B_k \varepsilon \mathcal{B}(R)$, let $B_m = R$ for $m = k + 1, \ldots, n$. By exchangeability and selection of π, it follows that

$$P[X_1 \ \varepsilon B_1, \ldots, X_k \varepsilon B_k] = P[\bigcap_{i=1}^{n} [X_i \varepsilon B_i]]$$

$$= P[\bigcap_{i=1}^{n} [X_{\pi i} \varepsilon B_i]]$$

$$= P[\bigcap_{i=1}^{k} [X_{\pi i} \varepsilon B_i]] = P[X_{i_1} \varepsilon B_1, \ldots, X_{i_k} \varepsilon B_k]. \ ///$$

An immediate consequence of Proposition 1.1.1 is that exchangeable random variables are stationary. However, for stationary random variables the dependence may vary between elements in the sequence which is not the case for exchangeable random variables. Thus, stationary random variables need not be exchangeable. The proof of Proposition 1.1.1 illustrates also that independent, identically distributed random variables are exchangeable. Example 1.1.1 will show that exchangeable random variables need not be independent and will serve as a generic example for exchangeability and what can go wrong when considering limit theorems.

Example 1.1.1 Let $\{Y_n\}$ be a sequence of independent, identically distributed random variables and let Y be a random variable independent of the sequence $\{Y_n\}$. Then by letting $X_n = Y_n + Y$, it can easily be shown that $\{X_n\}$ is a sequence of exchangeable random variables. Next, assuming $EY_n = 0 = EY$ for each n, $\text{Var } X_n = E(Y_n + Y)^2 = E(Y_n^2) + 2E(Y_nY) + E(Y^2) = \text{Var } Y_1 + 0 + \text{Var } Y$ for each n by identical distributions. Moreover, all joint distributions of the sequence $\{X_n\}$ are the same which implies a common correlation ρ^* for the sequence $\{X_n\}$. In particular,

$$\rho^* = \frac{E(X_1 X_2) - (EX_1)(EX_2)}{\sqrt{(\text{Var } X_1)(\text{Var } X_2)}}$$

$$= \frac{E(Y_1Y_2 + YY_1 + YY_2 + Y^2) - (EY_1 + EY)(EY_2 + EY)}{\text{Var } X_1}$$

$$= \frac{E(Y^2) - (EY)^2}{\text{Var } Y_1 + \text{Var } Y} = \frac{\text{Var } Y}{\text{Var } Y_1 + \text{Var } Y} . \qquad (1.1.2)$$

Thus, if $\text{Var } Y \neq 0$, then $\{X_n\}$ are not independent. ///

Equation (1.1.2) motivates consideration of the bounds for the common coefficient ρ^*. Let $\{X_1, \ldots, X_n\}$ be exchangeable random variables with finite (common) variance $\sigma^2 > 0$. Then

$$0 \leq \text{Var}(\Sigma_{i=1}^n X_i) = \Sigma_{i=1}^n \text{Var } X_i + \sum_{i \neq j}\sum \text{Cov}(X_i, X_j)$$

$$= n\sigma^2 + n(n-1)\rho^* \sqrt{\sigma^2 \ \sigma^2}$$

$$= \sigma^2 n(1 + (n-1)\rho^*). \qquad (1.1.3)$$

Dividing by $\sigma^2 n$ in Inequality (1.1.3) yields

$$\frac{-1}{n-1} \leq \rho^* \leq 1. \qquad (1.1.4)$$

Inequality (1.1.4) shows that an infinite sequence of exchangeable random variables can not be negatively correlated. In fact, infinite exchangeability implies much more than finite exchangeability as will be illustrated by de Finetti's theorem in Section 1.3. Example 1.1.2 will show that several interesting examples of negatively correlated exchangeable random variables exist.

Example 1.1.2 Let $A_1, \ldots, A_N$ be N distinct real numbers. Let X_i denote the ith value selected by random sampling without replacement, from $A_1, \ldots, A_N$ $1 \leq i \leq n$ where $n \leq N$. Then $\{X_1, \ldots, X_n\}$ are exchangeable, and

$$P[X_j = A_i] = \frac{1}{N} \quad \text{for all } i,j \tag{1.1.5}$$

$$\mathrm{Var}(X_j) = \frac{1}{N^2} (N\Sigma_{i=1}^N A_i^2 - (\Sigma_{i=1}^N A_i)^2) \tag{1.1.6}$$

$$\mathrm{Cov}(X_i, X_j) = \frac{1}{N^2} \frac{-1}{N-1} (N\Sigma_{i=1}^N A_i^2 - (\Sigma_{i=1}^N A_i)^2). \tag{1.1.7}$$

From (1.1.6) and (1.1.7) it follows that $\rho^* = \frac{-1}{N-1}$. ///

Since many test statistics in nonparametric statistics are sums of ranks or similar exchangeable random variables, Example 1.1.2 illustrates the importance of considering convergence of weighted sums or averages of exchangeable random variables. For another example, let $\{Y_1, \ldots, Y_n\}$ be independent, identically distributed random variables. Define $X_i = Y_i - \bar{Y}$ where $\bar{Y} = 1/n(Y_1 + \ldots + Y_n)$. It can be shown (using characteristic functions) that $\{X_1, \ldots, X_n\}$ are exchangeable. Moreover, the following property, Property 1.1.2, using $h(x) = x^\alpha$, $\alpha > 0$, can establish that $\{X_1^\alpha, \ldots, X_n^\alpha\}$ are also exchangeable. Thus, the centered moments $\frac{1}{n} \Sigma_{i=1}^n (Y_i - \bar{Y})^\alpha$ are also averages of exchangeable random variables.

Property 1.1.2 Let $\{X_n\}$ be exchangeable random variables. If $h : R \to R$ is a Borel function, then $h(X_1), \ldots, h(X_n)$ are exchangeable.

Proof For any permutation π of $\{1,2,\ldots,n\}$ and Borel subsets $B_1, \ldots, B_n$,

$$P[\bigcap_{i=1}^{n} [h(X_{\pi i}) \in B_i]] = P[\bigcap_{i=1}^{n} [X_{\pi i} \in h^{-1}(B_i)]].$$

By exchangeability,

$$P[\bigcap_{i=1}^{n} [X_{\pi i} \in h^{-1}(B_i)]] = P[\bigcap_{i=1}^{n} [X_i \in h^{-1}(B_i)]]$$

$$= P[\bigcap_{i=1}^{n} [h(X_i) \in B_i]].$$

Hence, it follows that $h(X_i), \ldots, h(X_n)$ are exchangeable.///

1.2 EXCHANGEABILITY AND LIMIT THEOREMS

In the previous section several examples of exchangeable random variables were given, and the importance of considering limits of weighted sums or averages of exchangeable random variables was illustrated. In considering stochastic convergence results, it is interesting to note that the absolute moments of the averages of exchangeable random variables can not increase (Exercise 10, Chow and Teicher (1978), p. 227).

Proposition 1.2.1 If $\{X_n\}$ are exchangeable random variables and $E|X_1| < \infty$, then for each m

$$E|\frac{1}{m} \Sigma_{k=1}^{m} X_k| \le E|\frac{1}{m-1} \Sigma_{k=1}^{m-1} X_k|.$$

Proof: Define a sequence of σ-fields by $F_m = \sigma\{\sum_{i=1}^{m} X_i, X_{m+1}, X_{m+2}, \ldots\}$. From Property 1.2.4 (to be established later), it follows that

$$\frac{1}{m}\Sigma_{k=1}^{m} X_k = E(X_1|F_m) \quad \text{a.s.} \tag{1.2.1}$$

Since $F_m \subset F_{m-1}$ for each m,

$$\begin{aligned}|\tfrac{1}{m}\Sigma_{k=1}^{m} X_k| &= |E(X_1|F_m)| \\ &= |E(E(X_1|F_{m-1})|F_m)| \\ &= |E(\tfrac{1}{m-1}\Sigma_{k=1}^{m-1} X_k|F_m)| \quad \text{a.s.} \end{aligned} \tag{1.2.2}$$

from (1.2.1). Taking expectations of both sides of (1.2.2) yields

$$\begin{aligned}E|\tfrac{1}{m}\Sigma_{k=1}^{m} X_k| &= E|E(\tfrac{1}{m-1}\Sigma_{k=1}^{m-1} X_k|F_m)| \\ &\le E(E(|\tfrac{1}{m-1}\Sigma_{k=1}^{m-1} X_k||F_m)) \\ &= E|\tfrac{1}{m-1}\Sigma_{k=1}^{m-1} X_k|. \end{aligned}$$

///

Remark: A shorter proof using martingales is suggested by Problem 1.9.

It is important to note that Proposition 1.2.1 holds for both infinite and finite exchangeable sequences. However, Example 1.1.1 shows that the limit of the averages of a sequence of exchangeable random variables need not be a constant since

$$\begin{aligned}\frac{1}{n}\Sigma_{k=1}^{n} X_k &= \frac{1}{n}\Sigma_{k=1}^{n} Y_k + \frac{1}{n}\Sigma_{k=1}^{n} Y \\ &\to 0 + Y \quad \text{a.s.}\end{aligned}$$

when $E|Y_1| < \infty$ and $EY_1 = 0$. In particular, the limit of $\frac{1}{n}\sum_{k=1}^{n} X_k$ must be measurable with respect to the tail σ-field, $\mathcal{T} = \bigcap_{n=1}^{\infty} \sigma\{X_n, X_{n+1}, \ldots\}$, which is essentially (including sets of probability 0 and 1) $\sigma\{Y\}$ since $\bigcap_{n=1}^{\infty} \sigma\{Y_n, Y_{n+1}, \ldots\}$ contains only sets of probability 0 and 1 by Kolmogorov's zero-one law. Possible limits for weighted sums of exchangeable random variables are better suggested by the following definitions and properties on conditional independence and exchangeability from Chow and Teicher (1978).

Definition 1.2.1 Let $\mathcal{G}$ be a σ-field and $\{\mathcal{G}_n\}$ be a sequence of classes of events. The sequence $\{\mathcal{G}_n\}$ is said to be conditionally independent given $\mathcal{G}$ if for each m = 1, 2, ..., n with $n \geq 2$ and for each $A_m \in \mathcal{G}_{k_m}$, $k_i \neq k_j$ when $i \neq j$,

$$P\left[\bigcap_{i=1}^{n} A_i \,|\, \mathcal{G}\right] = \prod_{i=1}^{n} P[A_i \,|\, \mathcal{G}] \quad \text{a.s.}$$

A sequence of random variables $\{X_n\}$ is said to be conditionally independent given $\mathcal{G}$ if the sequence of classes $\{\sigma(X_n)\}$ is conditionally independent given $\mathcal{G}$.

Let T denote the set of finite permutations on the positive integers, $\mathcal{B}^\infty$ be the Borel subsets of $R^\infty = R \times R \times \ldots$, and $X = (X_1, X_2, \ldots)$. The σ-field of permutable events (of X) is defined by

$$\mathcal{C} = \{X^{-1}(B) : B \in \mathcal{B}^\infty,\ P[X^{-1}(B) \,\Delta\, (\pi X)^{-1}(B)] = 0,\ \pi \in T\}.$$

where $\pi X = (X_{\pi 1}, X_{\pi 2}, \ldots)$ and Δ denotes the symmetric difference of two sets. Also, the permutable σ-field of order n can be defined by

$$\mathcal{C}_n = \{X^{-1}(B) : B \in \mathcal{B}(R^n),\ P[X^{-1}(B) \,\Delta\, (\pi X)^{-1}(B)] = 0 \text{ for all permutations } \pi \text{ which permutes at most the first n indices}\}$$

with $X = (X_1, \ldots, X_n)$ and $\pi X = (X_{\pi 1}, \ldots, X_{\pi n})$.

In addition, $C = \bigcap_{n=1}^{\infty} C_n$.

Theorem 1.2.2 (Chow and Teicher (1978), p. 220). Random variables $\{X_n : n \geq 1\}$ are exchangeable if and only if they are conditionally independent and identically distributed given some σ-field of events, G. Moreover, G can be taken to be either the σ-field C of permutable events or the tail σ-field T, and

$$P[X_1 \leq \alpha | C] = P[X_1 \leq \alpha | T] \quad \text{a.s.} \tag{1.2.3}$$

for each $\alpha \in R$.

Proof: The "if" part follows easily since

$$P[X_1 \leq x_1, \ldots, X_m \leq x_m | G] = \prod_{j=1}^{m} P[X_1 \leq x_j | G]$$

is invariant with respect to permutations and exchangeability can be established by taking expectations. To prove the "only if" part, consider the following: For $n \geq 1$ and any real number x, define

$$\xi_n(x) = \frac{1}{n} \sum_{i=1}^{n} I_{[X_i \leq x]}. \tag{1.2.4}$$

Then

$$E[\xi_n(x) - \xi_m(x)]^2 = \frac{|m-n|}{m \cdot n} (P[X_1 \leq x] - P[X_1 \leq x, X_2 \leq x])$$

$$\longrightarrow 0 \text{ as } m, n \longrightarrow \infty,$$

whence $\xi_n(x) \overset{p}{\rightarrow}$ to some T measurable function $\xi(x)$. Since $\xi_n(x) \overset{p}{\longrightarrow} \xi(x)$, it follows that $g(\xi_n(x)) \overset{p}{\longrightarrow} g(\xi(x))$ for each continuous function g. In particular, for $g(x) = \ln(x)$ $\sum_{j=1}^{m} \ln \xi_n(x_j) \overset{p}{\longrightarrow} \sum_{j=1}^{m} \ln \xi(x_j)$. This implies that $\ln \prod_{j=1}^{m} \xi_n(x_j) \xrightarrow{p} \sum_{j=1}^{m} \ln \xi(x_j)$. Hence it follows that

$$\prod_{j=1}^{m} \xi_n(x_j) \xrightarrow{p} \prod_{j=1}^{m} \xi(x_j), \quad m \geq 1. \tag{1.2.5}$$

Let $\alpha_1, \ldots, \alpha_m$ denote positive integers and set

$M = \{\alpha : \alpha = (\alpha_1, \ldots, \alpha_m)$, where $1 \leq \alpha_i \leq n$ for $1 \leq i \leq m\}$

$N = \{\alpha : \alpha \in M, \alpha_i \neq \alpha_j, i \neq j, 1 \leq i, j \leq m\}$.

$I_\alpha = I_{[X_{\alpha_1} \leq x_1, \ldots, X_{\alpha_m} \leq x_m]}$.

Whenever $n > m$, taking cognizance of (1.2.6),

$$\prod_{j=1}^{m} \xi_n(x_j) = \prod_{j=1}^{m} \left(\frac{1}{n} \sum_{i=1}^{n} I_{[X_i \leq x_j]}\right)$$

$$= \frac{1}{n^m} \prod_{i=1}^{m} \sum_{i=1}^{n} I_{[X_i \leq x_j]}$$

$$= \frac{1}{n^m} \sum_{\alpha \in M} I_\alpha$$

$$= \frac{1}{n^m} \left(\sum_{\alpha \in M-N} I_\alpha + \sum_{\alpha \in N} I_\alpha \right). \tag{1.2.7}$$

Note that M contains n^m terms and N contains $\frac{n!}{(n-m)!}$ terms. Thus, as $n \to \infty$

$$\frac{1}{n^m} \sum_{\alpha \in M-N} I_\alpha \leq \frac{1}{n^m} \sum_{\alpha \in M-N} 1 = \frac{1}{n^m} \left(n^m - \frac{n!}{(n-m)!}\right)$$

$$= \frac{n^m - (n(n-1) \ldots (n-m+1))}{n^m} \to 0$$

for each element $A = X^{-1}(B) \in C$, (1.2.7) entails

$$\int_A \prod_{j=1} \xi_n(x_j)\, dP = n^{-m} \sum_{\alpha \in N} \int_{X^{-1}(B)} I_\alpha\, dP + o(1) \quad (1.2.8)$$

[where $X = (X_1, X_2, \ldots)$ and Q = set of all finite permutations π from the set of all positive integers N into itself where π is one-to-one and $\pi n = n$ for all but a finite number of integers]. Thus, for any $\alpha \varepsilon N$, if $\pi \varepsilon Q$ is chosen so that $\pi i = \alpha_i$, $1 \leq \alpha \leq m$, then by definition of C and exchangeability

$$\int_{X^{-1}(B)} I_\alpha\, dP = \int_{(\pi X)^{-1}(B)} I_{[X_{\alpha_1} \leq x_1, \ldots, X_{\alpha_m} \leq x_m]}\, dP$$

$$= \int_{X^{-1}(B)} I_{[X_1 \leq x_1, \ldots, X_m \leq x_m]}\, dP.$$

Thus,

$$\int_A \prod_{j=1}^{m} \xi_n(x_j)\, dP = \frac{n(n-1)\ldots(n-m+1)}{n^m} \int_{X^{-1}(B)} I_{[X_1 \leq x_1, \ldots, X_m \leq x_m]}\, dP + o(1)$$

$$\longrightarrow \int_A I_{[X_1 \leq x_1, \ldots, X_m \leq x_m]}\, dP. \quad (1.2.9)$$

On the other hand, the left side of (1.2.9) tends to $\int_A \prod_{j=1}^{m} \xi(x_j)\, dP$ by (1.2.4) and the dominated congergence theorem. Hence, for any $A \varepsilon C$, all real $x_1, \ldots, x_m$, and positive integers m, $\int_A \prod_{j=1}^{m} \xi(x_j)\, dP = \int_A I_{[X_1 \leq x_1, \ldots, X_m \leq x_m]}\, dP$

and therefore

$$E[\prod_{j=1}^{m} \xi(x_j) | C] = E[I_{[X_1 \leq x_1, \ldots, X_m \leq x_m]} | C]$$

$$= P[\ _1 \leq x_1, \ldots, X_m \leq x_m | C] \text{ a.s.} \quad (1.2.10)$$

Since $\mathcal{C} \supset \mathcal{T}$ and $\xi(x)$ is T-measurable for every x, (1.2.9) entails

$$P[X_1 \le x_1, \ldots, X_m \le x_m | \mathcal{C}] = \prod_{i=1}^{m} \xi(x_j). \quad \text{a.s.} \tag{1.2.11}$$

If $A = X^{-1}(B) \varepsilon T$, it follows that

$$P[X_1 \le x_1, \ldots, X_m \le x_m \mid \mathcal{T}] = E[\prod_{j=1}^{m} \xi(x_j) \mid \mathcal{T}]$$
$$= \prod_{j=1}^{m} \xi(x_j). \quad \text{a.s.} \tag{1.2.12}$$

In particular, for m = 1,

$\xi(x_1) = P[X_1 \le x_1 \mid \mathcal{C}] = P[X_1 \le x_1 \mid \mathcal{T}]$ a.s. and by (1.2.10) and (1.2.11)

$$P[X_1 \le x_1, \ldots, X_m \le x_m \mid \mathcal{C}] = \prod_{i=1}^{m} P[X_1 \le x_i \mid \mathcal{C}]$$
$$= \prod_{i=1}^{m} P[X_1 \le x_i | \mathcal{T}]$$
$$= P[X_1 \le x_1, \ldots, X_m \le x_m \mid \mathcal{T}] \quad \text{a.s.} \tag{1.2.13}$$

Finally for any j = 2, ..., m, letting $x_i \to \infty$ for $i \ne j$, $1 \le i \le m$ in (1.2.13)

$$P[X_j \le x_j \mid \mathcal{C}] = P[X_1 \le x_j \mid \mathcal{C}]$$
$$= P[X_1 \le x_j \mid \mathcal{T}]$$
$$= P[X_j \le x_j \mid \mathcal{T}] \quad \text{a.s.} \tag{1.2.14}$$

Together (1.2.12) and (1.2.13) assert that $\{X_n : n \ge 1\}$ are conditionally i.i.d. and that (1.2.3) holds. ///

The "if" part of Theorem 1.2.2 holds for a finite sequence of random variables; however, infinite exchangebility is used crucially in the proof of the "only if" part. An immediate corollary of Theorem 1.2.2 is a form of de Finetti's Theorem.

Corollary 1.2.3 (de Finetti's Theorem): If $\{X_n : n \geq 1\}$ are exchangeable random variables, there exists a σ-field G such that

$$P[X_1 \leq \alpha_1, \ldots, X_m \leq \alpha_m] = \int \prod_{j=1}^{m} P[X_1 \leq \alpha_j | G] dP \qquad (1.2.15)$$

for all $\alpha_1, \ldots, \alpha_m \in R$.

Proof: Since the $\{X_n : n \geq 1\}$ are exchangeable, by Theorem 1.2.2, there exists a σ-algebra $G \subset A$ such that

$$P[X_1 \leq \alpha_1, \ldots, X_m \leq \alpha_m | G] = \prod_{j=1}^{m} P[X_1 \leq \alpha_j | G].$$

Thus,

$$P[X_1 \leq \alpha_1, \ldots, X_m \leq \alpha_m]$$

$$= \int P[X_1 \leq \alpha_1, \ldots, X_m \leq \alpha_m | G] dP$$

$$= \int \prod_{j=1}^{m} P[X_1 \leq \alpha_j | G] dP. \qquad ///$$

A popular expression of de Finetti's Theorem is that a sequence of exchangeable random variables is a mixture of sequences of independent, identically distributed random variables. More details on de Finetti's Theorem is provided in the next section.

Property 1.2.4 Let $\{X_n\}$ be exchangeable random variables and let $f : R \to R$ be a Borel function such that $E|f(X_1)| < \infty$. If $F_m = C_m$ (the permutable σ-field of order m) or

$F_m = \sigma\{\Sigma_{k=1}^m X_k, X_{m+1}, X_{m+2}, \ldots\}$ and $\sigma(\Sigma_{k=1}^m f(X_k)) \subset F_m$,

then

$$\frac{1}{m}\,\Sigma_{k=1}^{m} f(X_k) = E[f(X_1)|\mathcal{F}_m] \quad \text{a.s.}$$

<u>Proof</u>: Let $Y = g(X_1, X_2, \ldots)$ be a bounded random variable where g is a Borel function and $g(X_1, X_2, \ldots) = g(X_{\pi 1}, X_{\pi 2}, \ldots)$ a.s. for each permutation π of the first m indices. Then $h(x_1, x_2, \ldots) = f(x_1)g(x_1, x_2, \ldots)$ is a Borel function, and by exchangeability.

$$\begin{aligned} F_{f(X_j)Y}(\alpha) &= P[h(X_j, X_2, \ldots, X_{j-1}, X_1, X_{j+1}, \ldots) \le \alpha] \\ &= P[(X_j, X_2, \ldots, X_{j-1}, X_1, X_{j+1}, \ldots) \varepsilon\, h^{-1}(-\infty, \alpha]] \\ &= P[(X_1, X_2, \ldots, X_{j-1}, X_j, X_{j+1} \ldots) \varepsilon\, h^{-1}(-\infty, \alpha]] \\ &= P[h(X_1, X_2, \ldots, X_{j-1}, X_j, X_{j+1}, \ldots) \le \alpha] \\ &= F_{f(X_1)Y}(\alpha) \qquad (1.2.16) \end{aligned}$$

for each $1 \le j \le m$ and $\alpha \varepsilon R$. In particular, $E[f(X_j)Y] = E[f(X_1)Y]$ for each $1 \le j \le m$. When $Y = I_A$ with $A \varepsilon \mathcal{F}_m$, (1.2.3) provides that

$$\begin{aligned} \int_A \frac{1}{m}\,\Sigma_{k=1}^{m} f(X_k)dP &= \frac{1}{m}\,\Sigma_{k=1}^{m}\, E(f(X_k)I_A) \\ &= E(f(X_1)I_A) \\ &= \int_A f(X_1)dP \\ &= \int_A E(f(X_1)|\mathcal{F}_m)dP. \qquad (1.2.17) \end{aligned}$$

Hence,

$$\frac{1}{m}\sum_{k=1}^{m} f(X_k) = E(f(X_1)|F_m) \quad \text{a.s.} \qquad ///$$

Property 1.2.4 and its proof are from Kingman (1978) and do not require an infinite sequence of exchangeable random variables. Also, different choices of F_m are possible as long as (1.2.3) and (1.2.16) hold. Since $F_m \supset F_{m+1}$ for all m, $\{E(f(X_1)|F_m) : m \geq 1\}$ is a reverse martingale. When $\{X_n\}$ is an infinite sequence of exchangeable random variables, the Martingale Convergence Theorem (rf. Tucker (1967), p. 230) provides that $E(f(X_1)|F_m) \to E(f(X_1)|F_\infty)$ a.s. where $F_\infty = \bigcap_{m=1}^{\infty} F_m$. If $f(X_1) = X_1$, then

$$\frac{1}{m}\sum_{k=1}^{m} X_k = E(X_1|F_m) \to E(X_1|F_\infty) \quad \text{a.s.}$$

However, Example 1.1.1 shows that $E(X_1|F_\infty)$ need not be a constant as is the case for i.i.d. random variables. For convergence to a constant, correlation coefficient $\rho^* = 0$ is suggested for the laws of large numbers. In considering a central limit theorem,

$$\frac{1}{\sqrt{m}}\sum_{k=1}^{m} X_k = \sqrt{m}\, E(X_1|F_m),$$

it is again essential that $E(X_1|F_\infty)$ not be variable. However, rather than restricting the study of limit laws for exchangeable random variables to uncorrelated random variables, Example 1.2.1 suggests a more applicable format. Before considering Example 1.2.1, let $f(X_1) = I_{[X_1 \leq \alpha]}$ in Property 1.2.4 for some fixed $\alpha \in R$. Then the empirical distribution function for each $\alpha \in R$.

$$F_n(\alpha) = \frac{1}{n}\sum_{k=1}^{n} I_{[X_k \leq \alpha]} = E(I_{[X_1 \leq \alpha]}|F_n)$$

$$\overset{\text{a.s.}}{\to} E(I_{[X_1 \leq \alpha]}|F_\infty) = F_{X_1}(\alpha|F_\infty),$$

and a form of the Glivenko-Cantelli Theorem is available for exchangeable random variables.

Example 1.2.1: Let $\{X_{nk} : 1 \leq k \leq n, n \geq 1\}$ be an array of random variables such that for each n $\{X_{nk} : 1 \leq k \leq n\}$ are jointly normal with mean 0 and variance 1 (N(0,1)) and common correlation coefficient ρ_n^*. Thus, $\{X_{n1}, \ldots, X_{nn}\}$ are exchangeable, and $\Sigma_{k=1}^n X_{nk}$ is distributed as $N(0, n + n(n-1)\rho_n^*)$. For this example $\frac{1}{n}\Sigma_{k=1}^n X_{nk} \to 0$ in probability if and only if $\rho_n^* \to 0$ since $\frac{1}{n}\Sigma_{k=1}^n X_{nk}$ converges in distribution to a $N(0,\rho)$ random variable when $\rho_n^* \to \rho$. The condition $\rho_n^* \to 0$ will be frequently used in considering the laws of large numbers in Chapter III. For the central limit theorem to hold, $(n-1)\rho_n^*$ must go to zero ($\rho_n^* = o(\frac{1}{n})$). This condition will be used crucially in considering central limit theorems in Chapter II.

1.3 de FINETTI THEOREM AND MEASURABILITY PROPERTIES

Theorem 1.2.2 showed that an infinite sequence $\{X_n\}$ of exchangeable random variables was conditionally i.i.d. and obtained as a corollary a form of de Finetti's theorem

$$P[X_1 \leq \alpha_1, \ldots, X_m \leq \alpha_m] = \int \prod_{j=1}^{m} P[X_1 \leq \alpha_j|G]dP \qquad (1.3.1)$$

for all $\alpha_1, \ldots, \alpha_m \in R$. A particular form of de Finetti's theorem will be established in this section which will aid in calculating probabilities. In addition, several measurability properties will also be established in this section.

Let F be the collection of probability measures (or equivalently, the distribution functions) on $B(R)$. Let F be equipped with the σ-field generated by the topology of weak convergence of the distribution functions, that is, the σ-field is generated by sets of the form $\{F \in F: F(x) \leq y\}$ with x, y ε R. Doob (1940) showed that there exists a regular conditional distribution for $X_1, \ldots, X_m$ given a σ-field G. Thus, from (1.3.1) $\prod_{j=1}^{m} P[X_1 \leq \alpha_j | G]$ determines a probability on $\mathcal{B}(R^m)$ for almost every $w \in \Omega$. Hence, for any Borel function $g : R^m \to R$ and $B \in \mathcal{B}(R)$

$$P[g(X_1, \ldots, X_m) \in B] = \int P[g(X_1, \ldots, X_m) \in B|G]dP$$

$$= \int_F P_F(B)\mu(dF) \qquad (1.3.2)$$

where μ is a probability measure (termed the "mixing measure") on the prescribed σ-field on F. Combining (1.3.1) and (1.3.2) yields that $P_F(B)$ is calculated as if $X_1, \ldots, X_m$ are independent. Equation (1.3.2) is often expressed imprecisely as

$$P(A) = \int_F P_F(A)\mu(dF) \qquad (1.3.3)$$

where $A \in \sigma\{X_1, X_2, \ldots\}$ and $P_F(A)$ is understood to be $P[(X_1, X_2, \ldots) \in B|G]$ with $A = (X_1, X_2, \ldots)^{-1}(B)$ since the probabilities of these sets are defined by extending the measures. From (1.3.2)

$$E[g(X_1, \ldots, X_m)] = E[E_F[g(X_1, \ldots, X_m)]]$$
$$= E[\int_{-\infty}^{\infty} \ldots \int_{-\infty}^{\infty} g(x_1, \ldots, x_m)dF(x_1)\ldots dF(x_m)] \qquad (1.3.4)$$

In particular, the key orthogonality consideration is

$$E[X_1X_2] = E[\int_{-\infty}^{\infty} \int_{-\infty}^{\infty} x_1x_2 dF(x_1)dF(x_2)]$$

$$= E[(\int_{-\infty}^{\infty} xdF(x))^2] = E[(E_F X_1)^2]. \qquad (1.3.5)$$

Before addressing additional properties, it is important to observe in an example from Diaconis and Freedman (1980) the necessity of infinite exchangeability. Let

(X_1, X_2) have the joint probability distribution specified by

$$P[X_1 = 0,\ X_2 = 1] = \frac{1}{2} = P[X_1 = 1,\ X_2 = 0].$$

If a mixing measure μ existed, then

$$1 = P[(X_1,X_2) \varepsilon \{(0,1),\ (1,0)\}]$$

$$= \int_F P_F[X_1{=}0]P_F[X_2{=}1]\mu(dF) + \int_F P_F[X_1{=}1]P_F[X_2{=}0]\mu(dF)$$

$$= 2\int_F P_F[X_1 = 1]P_F[X_1 = 0]\mu(dF). \qquad (1.3.6)$$

But,

$$0 = P[X_1 = 1,\ X_2 = 1] = \int_F (P_F[X_1 = 1])^2\mu(dF)$$

and

$$0 = P[X_1 = 0,\ X_2 = 0] = \int_F (P_F[X_1 = 0])^2\mu(dF)$$

which implies that $P_F[X_1 = 0] = 0 = P_F[X_1 = 1]$ μ a.s. which contradicts (1.3.6). Moreover, this distribution has the very simple and practical interpretation of sampling without replacement from an urn with one white and one black ball.

This example extends to a finite number of random variables, but $\lim_{n\to\infty} (P_F[X_1 = 1])^n$ and $\lim_{n\to\infty} (P_F[X_1 = 0])^n$ can both be 0 without $P_F[X_1 = 0] = 0 = P_F[X_1 = 1]$ μ a.s. The example illustrates some Diaconis and Freedman (1980) results for finite approximate forms of deFinetti's theorem. That is, if a finite exchangeable sequence of random variables of length k assuming only the values 0 and 1 can be embedded in an exchangeable sequence of length $n > k$, then $P[X_1 = i_1, \ldots, X_k = i_k]$ is within $4k/n$ of an integral mixture of sequences of i.i.d. Bernoulli random variables for $i_1, \ldots, i_k \in \{0,1\}$. For k arbitrary random variables which can be embedded in an exchangeable sequence of length $n > k$, the bound is $k(k-1)/n$ for the difference between the probability of the induced measure on R^k by the first k variables of the exchangeable sequence and by a mixture of sequences of i.i.d. random variables.

For i.i.d. random variables $\{X_n\}$, the strong law of large numbers holds if and only if $E|X_1| < \infty$. The following example shows that the situation is quite different for exchangeable random variables.

<u>Example 1.3.1</u> Let $\{X_n : n \geq 1\}$ be a sequence of exchangeable random variables where the corresponding de Finetti mixture measure μ is concentrated on the distribution functions F_m with probability $\frac{1}{2^m}$, $m \geq 1$. Moreover, assume that $P_{F_m}[X_1 = 2^m] = \frac{1}{2} = P_{F_m}[X_1 = -2^m]$. For each F_m the SLLN holds, that is,

$$\lim_{\ell\to\infty} P_{F_m}[\bigcup_{n=\ell}^{\infty}[|\frac{1}{n}\Sigma_{k=1}^{n} X_k| > \varepsilon]] = 0$$

for each $\varepsilon > 0$. Thus, by the bounded convergence theorem,

$$\lim_{\ell\to\infty} P[\bigcup_{n=\ell}^{\infty}[|\frac{1}{n}\Sigma_{k=1}^{n} X_k| > \varepsilon]]$$

$$= \int_F \lim_{\ell\to\infty} P_F[\bigcup_{n=\ell}^{\infty}[|\frac{1}{n}\Sigma_{k=1}^{n} X_k| > \varepsilon]]d\mu(F)$$

$$= 0,$$

and hence $\frac{1}{n}\Sigma_{k=1}^{n} X_k \to 0$ a.s.

But,

$$E|X_1| = \int_0^{\infty} P[|X_1| > t]dt$$

$$= \int_0^\infty \int_F P_F[|X_1| > t]d\mu(F)dt$$

$$= \int_0^\infty \Sigma_{m=1}^\infty P_{F_m}[|X_1| > t]\frac{1}{2^m}\, dt$$

$$= \Sigma_{m=1}^\infty \int_0^\infty P_{F_m}[|X_1| > t]dt\frac{1}{2^m} = \Sigma_{m=1}^\infty 2^m \cdot \frac{1}{2^m} = \infty. \;///$$

A slight modification of Example 1.3.1 shows that the strong law of large numbers can hold and $E|X_1|^\alpha = \infty$ for $\alpha > 0$. Example 1.3.1 illustrates the importance of considering the conditional means $\{E_F X_1 : F \in F\}$ in obtaining limit theorems. The following proposition insures measurability of subsets of F determined by the conditional means.

Proposition 1.3.1 Let $\{X_n : n \geq 1\}$ be exchangeable random variables. For each $c > 0$ and $\eta > 0$

$$\{F \in F : |\, E_F(X_1 I_{[|X_1| \leq c]})\, | > \eta\} \tag{1.3.6}$$

is in the σ-field generated by the topology of weak convergence of distribution functions.

Proof: Let $g: R \to R$ be a continuous function with compact support. The mapping $\psi(F) = \int xg(x)dF(x)$ of $F \to R$ is continuous since $xg(x)$ is a bounded, uniformly continuous function and $F_n \to F$ if and only if $\int f(x)dF_n(x) \to \int f(x)dF(x)$ for every bounded, uniformly continuous function (rf: Theorem 4.2). For each $m \geq 1$, let g_m be the uniformly continuous function defined by $g_m(x) = 1$ if $|x| \leq c$, $g_m(x) = 0$ if $|x| \geq c + \frac{1}{m}$, and linear on both $(-c-\frac{1}{m}, -c)$ and $(c, c+\frac{1}{m})$. Note that $xg_m \to xI_{[|x| \leq c]}$ for each $x \in R$. By the bounded convergence theorem, $\psi_m(F) = \int xg_m(x)dF(x) \to \int xI_{[|x| \leq c]}dF = \psi_0(F)$ for each $F \in F$. Since $\psi_0(F)$ is the pointwise limit of a sequence of measurable functions, it is measurable with respect to the σ-field generated by the topology of weak convergence. Noting that

$$\{F \in \mathcal{F} : |E_F(XI_{[|X| \leq c]})| > \eta\} = \psi_0^{-1}((\eta,\infty))$$

and that $(\eta,\infty) \in \mathcal{B}(R)$ completes the proof. ///

<u>Corollary 1.3.2</u> If $\{X_n : n \geq 1\}$ are exchangeable random variables with $E|X_1| < \infty$, then

$$\{F \in \mathcal{F} : |E_F X_1| > \eta\} \tag{1.3.7}$$

is in the σ-field generated by the topology of weak convergence of distribution functions.

<u>Proof</u>: First, $E|X_1| = \int (E_F|X_1|)d\mu(F) < \infty$

implies that $E_F|X_1| < \infty$ μ a.s. In the proof of Proposition 1.3.1, let $\psi(F) = \lim_{m\to\infty} \int xI_{[|x| \leq m]}dF(x)$ which by the dominated convergence theorem is $\psi(F) = \int xdF(x) = E_F X_1$ and consequently is a measurable function from $\mathcal{F}$ into R. ///

The sets in (1.3.6) and (1.3.7) will play a role in obtaining limit laws for exchangeable random variables. Recall the format suggested by Example 1.2.1 that $\{X_{nk} : k \geq 1\}$ are exchangeable for each n and that common correlation coefficient ρ_n^* should go to 0 as $n \to \infty$. Assuming the moment condition $E|X_{n1}X_{n2}| < \infty$ is finite for each n, then

$$\rho_n = E(X_{n1}X_{n2}) = \int_{\mathcal{F}} E_F(X_{n1}X_{n2})\mu_n(dF) = \int_{\mathcal{F}} (E_F X_{n1})^2 \mu_n(dF)$$

$$\geq \int_{F_\eta} \eta^2 \mu_n(dF) = \eta^2 \mu_n(F_\eta) \qquad (1.3.8)$$

where $F_\eta = \{F \in F : |E_F X_{n1}| > \eta\}$ and μ_n is the mixing measure for each row $\{X_{nk} : k \geq 1\}$. Thus, ρ_n goes to 0 only if $\mu_n(F_\eta) \to 0$ for each η. Under certain bounded conditions on the random variables, the converse will also hold (see Problem 1.8). Also, ρ_n serves as measure of non-orthogonality, and $\rho_n \to 0$ seems (except for the changing μ_n's) to imply that the conditional means are going to 0. In Example 1.3.1 the conditional means $\{E_F X_1\}$ were 0 with μ-measure one but $\rho = EX_1X_2$ did not exist since $E|X_1X_2| = \infty$.

The last example shows that if $\{X_n : n \geq 1\}$ is a sequence of exchangeable random variables with $E(X_1X_2) = 0$ it need not follow that the tail σ-field contains only sets of probability zero or one. Let $\{Y_n\}$ be i.i.d. random variables with $P[Y_1 = 1] = 1/2 = P[Y_1 = -1]$ and let Z be a random variable independent of $\{Y_n\}$ with $P[Z = 1] = 1/2 = P[Z = 0]$. Let $X_n = Y_nZ$, and it is easy to see that $\{X_n\}$ are exchangeable. Note that $A = [Z = 0] \in \sigma\{Y_nZ, Y_{n+1}Z, \ldots\}$ for each n since $[Y_nZ = 0] = [Z = 0]$ a.s. ($P[Y_n = 0] = 0$ for all n). Thus, $A = [Z = 0] \in \bigcap_{n=1}^{\infty} \sigma\{X_n, X_{n+1}, \ldots\}$, but $P(A) = 1/2$.

1.4 PROBLEMS

1.1 Show that the random variables in Example 1.1.1 are exchangeable by using Definition 1.1.1 or Proposition 1.1.1.

1.2 Show that the random variables defined in Example 1.1.2 are exchangeable.

1.3 Verify (1.1.5), (1.1.6) and (1.1.7).

1.4 Use characteristic functions to show that $\{Y_i - \bar{Y} : 1 \leq i \leq n\}$ are exchangeable when $\{Y_1, \ldots, Y_n\}$ are independent, identically distributed random variables.

1.5 Show that $C = \bigcap_{n=1}^{\infty} C_n$.

1.6 In Example 1.2.1 show that the weak law of large numbers hold if and only if $\rho_n^* \to 0$.

1.7 Show that the σ-field generated by the topology of weak convergence of distribution functions is also generated by sets of the form $\{F \in \mathcal{F} : F(x) \leq y\}$ with $x, y \in R$.

1.8 Let $\{X_{nk} : k \geq 1\}$ be exchangeable random variables for each n which are uniformly bounded in n and k. Show that $\mu_n(\{F \in \mathcal{F} : |E_F X_{n1}| > \eta\}) \to 0$ for each $\eta > 0$ implies that $\rho_n = E(X_{n1} X_{n2}) \to 0$.

1.9 Use the fact that $\{E(X_1 | \mathcal{F}_m) = \frac{1}{m} \Sigma_{k=1}^m X_k : m \geq 1\}$ is a reverse martingale and hence $\{|E(X_1 | \mathcal{F}_m)| : m \geq 1\}$ is a reverse submartingale to prove Proposition 1.2.1.

1.10 With reference to Theorem 1.2.2. Show that

$$E[\xi_n(x) - \xi_m(x)]^2 = \frac{|m-n|}{mn} (P[X_1 \leq x] - P[X_1 \leq x, X_1 \leq x])$$

1.11 Prove statement (1.2.10) of the proof of Theorem 1.2.2.

CHAPTER II

CENTRAL LIMIT THEOREMS FOR EXCHANGEABLE RANDOM VARIABLES

2.0 INTRODUCTION

The investigation of central limit theorems for exchangeable random variables began with the apparently independent work of Blum et al. (1958) and Bühlmann (1958) & (1960). Both were concerned with infinite exchangeable sequences and both used as their basic point of departure the de Finetti integral representation (1.3.3). Both showed that for sequences $\{X_n\}$ of centered exchangeable r.v.s. with finite variance, the limit laws are weighted normal, and both obtained necessary and sufficient conditions for the limit law to be centered normal. The Raikov-type condition $n^{-1}\Sigma_1^n X_i^2 \overset{P}{\to} 0$ obtained by Bühlmann and the condition of uncorrelation of X_1^2 and X_2^2 of Blum et al. are closely related. Blum et al. (1958) obtained central limit theorems for double arrays with infinite exchangeable rows, again using the de Finetti representation. Chernoff and Teicher (1958) obtained central limit theorems for triangular arrays with exchangeable rows of finite length, under an additional condition that each row sum to a non-random constant, using ingenious techniques exploiting a theorem of Noether. A good reference for these type results is Chow and Teicher (1978).

Bühlmann (1960) investigated random processes with exchangeable increments, a field of study which was pursued in a series of papers by Kallenberg [see Kallenberg (1973), (1982) for further references] but which lack of space does not permit to be treated here. An invariance principle for exchangeable random variables yielding convergence in law to the familiar Brownian Bridge can be found in Chapter 4, Section 24, of Billingsley (1968).

It has long been known that Doob's reverse martingale proof of the strong law of large numbers for independent and identically distributed random variables carries over to an infinite exchangeable sequence (rf: Kingman (1978)). It was noticed by Weber (1980) that known martingale central limit theorems could be exploited to obtain central limit theorems for exchangeable random variables. Here de Finetti's theorem does not enter. Only the symmetry of the joint distributions is utilized, rendering the method suitable for dealing with triangular arrays with finite exchangeable rows, the lengths of which tend to infinity. (A fortiori, infinite exchangeable arrays can also be handled, yielding an alternative method for this case.) Weber (1980), using both direct and reverse martingale techniques, obtained central limit theorems and invariance principles yielding weak convergence to a normal distribution and Brownian motion for such arrays. The above-mentioned central limit theorem of Chernoff and Teicher (1958) and results very close to those of Blum et al. (1958) and Bühlmann (1958) & (1960) are obtained as corollaries.

The survey paper by Eagleson in the conference on exchangeability edited by Koch and Spizzichino (1982) presents results on weak convergence of exchangeable arrays (finite and infinite rows) to Poisson limits, discusses weak and partial exchangeability, and contains further references.

In Section 2.1 the de Finetti representation is used to obtain sufficient conditions for convergence in law of weighted sums of a sequence of exchangeable random variables to a limit law which is a mixture of normal distributions. For non-negative weights, the conditions are also necessary. The result of Blum et al. (1958) emerges as a corollary.

Section 2.2 deals with arrays. We have preferred to treat row-wise exchangeable arrays using martingale methods. In this way triangular arrays can be dealt with, and stronger results for arrays with exchangeable rows which are embeddable in infinite exchangeable sequences are available as corollaries. Specialization also yields central limit theorems for weighted sums of a single exchangeable sequence, which can be compared with those obtained using the de Finetti theorem.

2.1 CENTRAL LIMIT THEOREMS USING DE FINETTI'S THEOREM

In the case of an infinite exchangeable sequence or an array with rows which are infinitely exchangeable, the de Finetti representation allows results for independent random variables to be exploited. In this section a central limit theorem for an infinite exchangeable sequence will be proved and a now classical result derived as a corollary. We shall need the following result.

Lemma 2.1.1 If $\{X_n : n \geq 1\}$ is an exchangeable sequence of random variables with $E|X_1|^r < \infty$, where $r \geq 0$, then $E_F|X_1|^r = E|X_1|^r$ for μ-almost every $F \in \mathcal{F}$ (where μ is the de Finetti mixing measure for $\{X_n\}$) if and only if the random variables $\{|X_n|^r : n \geq 1\}$ are pairwise uncorrelated.

Proof: Put $m = E|X_1|^r$. Then $E[(|X_1|^r - m)(|X_2|^r - m)] = EE_F[(|X_1|^r - m)(|X_2|^r - m)] = E(E_F(|X_1|^r - m))^2 = 0$ if and only if $E_F|X_1|^r = m$, μ-a.s. ///

A probability distribution $F \in \mathcal{F}$ (or a r.v. Z with distribution function F) will be said to be a mixture of (centered) normals if $F(x) = \int_R \Phi(\sigma^{-1}x)dG(\sigma)$, where G is a distribution function concentrated on $[0,\infty)$, that is, $G(0-) = 0$. Here Φ is the normal cumulative, and for $\sigma = 0$ we put $\Phi(\sigma^{-1}x) = I_{[0,\infty)}(x)$ for $x \in R$.

To clarify the concept of a mixture of normals, let Z be a random variable with distribution function $F(x) = \int \Phi(\sigma^{-1}x)dG(\sigma)$. Then $Z \stackrel{d}{=} \eta N$, where η is a random variable with distribution function G, and N is a standard normal random variable, independent of η. Indeed (the integrals are all taken over R), $Ee^{itZ} = \int e^{itx}dF(x)$

$$= \int e^{itx} d \int \Phi(\sigma^{-1}x)dG(\sigma)$$

$$= \int e^{itx} \int \sigma^{-1}\phi(\sigma^{-1}x)dG(\sigma)dx$$

$$= \int \int e^{it\sigma x} \phi(x)dxdG(\sigma) = Ee^{it\eta N} .$$

Theorem 2.1.2 Let $X_1, X_2, \ldots$ be an infinite sequence of exchangeable random variables satisfying $EX_1 = 0$, $EX_1^2 = \sigma^2 < \infty$

and $EX_1X_2 = 0$. Let $\{a_{nk} : 1 \leq k \leq k_n, n \geq 1\}$ be an array of real constants and put $\bar{a}_n = \max_{1 \leq k \leq k_n} |a_{nk}|$ and $A_n = \Sigma_{k=1}^{k_n} a_{nk}^2$. Assume

(a) $\bar{a}_n \to 0$, and

(b) $A_n \to 1$, as $n \to \infty$.

Then

$$S_n = \Sigma_{k=1}^{k_n} a_{nk} X_k \xrightarrow{d} Z,$$

where Z is a random variable whose distribution is a mixture of normals.

Proof: De Finetti's theorem (1.3.3) allows us to write

$$P[S_n \leq x] = \int_F P_F[S_n \leq x]\, \mu(dF), \quad x \in R.$$

By Lemma 2.1.1 ($r = 1$), $E_F X_1 = 0$ for almost every F in the support of μ.

Put $X_{nk} = a_{nk}X_k$. By (a) $\{X_{nk} : 1 \leq k \leq k_n, n \geq 1\}$ is an infinitesimal array (that is, $\max_{1 \leq k \leq k_n} P_F[|X_{nk}| > \varepsilon] \to 0$, for every $\varepsilon > 0$). From (b) we get $\Sigma_{k=1}^{k_n} E_F X_{nk}^2 = A_n E_F X_1^2 \to E_F X_1^2 = \sigma_F^2$, for μ-almost every $F \in \mathcal{F}$. It follows from classical central limit theorems (see for example Chow & Teicher (1978), p.434, Corollary 2) that $P_F[\Sigma_{k=1}^{k_n} X_{nk} \leq x] \to \Phi(\sigma_F^{-1} x)$, for all $x \in R$, if and only if $\Sigma_{k=1}^{k_n} E_F X_{nk}^2 I_{[|X_{nk}| > \varepsilon]} \to 0$, for all $\varepsilon > 0$ (Lindeberg condition). But this is satisfied: $\Sigma_{k=1}^{k_n} a_{nk}^2 E_F X_1^2 I_{[|X_1| > \varepsilon \bar{a}_n^{-1}]} \to 0$ for μ-almost every F, using (a) and (b). Hence we get, using the bounded convergence theorem,

$$P[S_n \le x] = \int_F P_F[S_n \le x]\mu(dF)$$

$$\to \int_F \Phi(\sigma_F^{-1}x)\mu(dF), \quad \text{for all } x \in R, \text{ which can be}$$

written as $\int_{-\infty}^{\infty}\Phi(\sigma^{-1}x)dG(\sigma)$, where $G(\sigma) = \mu\{F \in \mathcal{F}: E_F X_1^2 \le \sigma^2\}$.///

Remark 2.1.1 If the weights $\{a_{nk}\}$ in Theorem 2.1.2 are non-negative, then a converse of the theorem is true: If $S_n \xrightarrow{d} Z$ then, necessarily, $EX_1X_2 = 0$, that is, the random variables of the sequence $\{X_k : k \ge 1\}$ are uncorrelated. The straight forward but rather technical demonstration of this fact is relegated to Problem 2.1.

Remark 2.1.2 The limit random variable Z in Theorem 2.1.2 is normally distributed if and only if $E_F X_1^2 = \sigma^2$, a constant, for μ-almost every F. According to Lemma 2.1.1 this happens if and only if X_1^2 and X_2^2 are uncorrelated [see Blum et al. (1958), or Chow & Teicher (1978), Section 9.2] If the mixing measure μ is concentrated on a family of probability distributions which is parametrized by its variance (for example, the centered normal, the Poisson, the exponential distributions, the binomial for fixed sample size), then uncorrelation of X_1^2 and X_2^2 implies that μ is concentrated on a single probability distribution and $X_1, X_2, \ldots$ are independent (and identically distributed) random variables.

Taking $a_{nk} = n^{-1/2}$, $k = 1, \ldots, n$ $(k_n = n)$ in Theorem 2.1.2 we obtain a result of Blum et al. (1958) [see also Chow & Teicher (1978), Section 9.2, Theorem 1].

Corollary 2.1.3 Let $\{X_n : n \ge 1\}$ be a sequence of exchangeable random variables satisfying $EX_1=0$ and $EX_1^2=1$.

Then $n^{-1/2} \Sigma_{k=1}^{n} X_k \to N(0,1)$ if and only if $EX_1X_2 = 0$ and $EX_1^2X_2^2 = 1$.

<u>Proof</u>: The "if" part follows directly from Theorem 2.1.2, Lemma 2.1.1 and Remark 2.1.2.

For the "only if" part, we first obtain from Remark 2.1.1 (Problem 2.1) that $EX_1X_2 = 0$. Since $P[n^{-1/2} \Sigma_{k=1}^{n} X_k \leq x] \to \Phi(x)$, and $\int_F P_F[n^{-1/2} \Sigma_{k=1}^{n} X_k \leq x]\mu(dF) \to \int_F \Phi(\sigma_F^{-1}x)\mu(dF)$ $= \int_R \Phi(\sigma^{-1}x)dG(\sigma)$, where $\sigma_F^2 = E_F X_1^2$, and G is a distribution function concentrated on $[0,\infty)$, we have $\int_{-\infty}^{\infty} \Phi(\sigma^{-1}x)dG(\sigma)$ $= \Phi(x)$, $\forall\, x \in R$. This implies that $G = I_{[1,\infty)}$, that is, G is concentrated at one (Problem 2.3). Thus, $\sigma_F = E_F X_1^2 = 1$ for μ-almost every F, and Lemma 2.1.1 yields $E(X_1^2 - 1)(X_2^2 - 1) = 0$, or $EX_1^2X_2^2 = 1$. ///

The practical value of a central limit theorem is enormously enhanced if one is able to say something about the rate of convergence of the normed sums to the limit distribution. For the case of the classical weights $a_{nk} = n^{-1/2}$, $k = 1, \ldots, n$, the celebrated Berry-Esséen result on the rate of convergence of normed sums of i.i.d. random variables to the standard normal distribution can be used in conjunction with the de Finetti representation to obtain rates of convergence for exchangeable random variables to mixtures of normals. For simplicity, we treat only the case of finite third moments.

If $\{X_n : n \geq 1\}$ is an i.i.d. sequence of random variables with $EX_1 = 0$, $\sigma^2 = EX_1^2$ and $\gamma^3 = E|X_1|^3 < \infty$, then with $S_n = X_1 + \ldots + X_n$, the Berry-Esséen theorem [cf. Chow and Teicher (1978) Corollary 4, p. 300] yields

$$\sup_{x\in R} |P[S_n \leq \sqrt{n}\,\sigma\, x] - \Phi(x)| \leq \frac{1.33}{\sqrt{n}} \frac{\gamma^3}{\sigma^3}$$

Now, supposing $\{X_n : n \geq 1\}$ to be a sequence of exchangeable random variables with $EX_1 = 0$ and $E|X_1|^3 < \infty$, and $n^{-1/2}S_n \xrightarrow{d} Z$, a mixture of normals. The de Finetti representation (1.3.3) then yields

$$\sup_{x\in R} |P[S_n \leq \sqrt{n}\, x] - F_Z(x)|$$

$$= \sup_{x\in R} \left| \int_F P_F[S_n \leq \sqrt{n}\, x]\mu(dF) - \int_F \Phi(\sigma_F^{-1}x)\mu(dF) \right| \leq$$

$$\leq \int_F \sup_{x\in R} |P_F[S_n \leq \sqrt{n}\, x] - \Phi(\sigma_F^{-1}x)|\mu(dF)$$

$$\leq \frac{1.33}{\sqrt{n}} \int_{F_+} \frac{\gamma_F^3}{\sigma_F^3}\mu(dF), \text{ where } \sigma_F^2 = E_F X_1^2 ,$$

$\gamma_F^3 = E_F|X_1|^3$ and $F_+ = \{F \in F : \int_R x^2 dF(x) > 0\}$.

Thus, if F^* denotes the σ-field of the topology of weak convergence on F, if $F_+^* = F_+ \cap F^*$, and putting $\phi(F) = (\gamma_F/\sigma_F)^3$, then what is needed for a rate of convergence is

$$\phi \in L^1((F_+, F_+^*, \mu); R), \text{ that is,}$$

$$\int_{F_+} \left(\frac{\gamma_F}{\sigma_F}\right)^3 \mu(dF) < \infty \qquad (2.1.1)$$

The integral (2.2.1) will need to be examined in special cases. In some cases, where the mixing measure μ is concentrated on a parametric family, the ratio γ_F/σ_F turns out to be a constant. This happens, for example, for the family of centered normal distributions $\{f_{\sigma^2}(x) = \frac{1}{\sqrt{2\pi}\sigma} \exp -\frac{x}{2\sigma^2} : \sigma^2 > 0\}$; for the family $\{f_\theta = \frac{1}{2\theta} I_{[-\theta,\theta]} : \theta > 0\}$ of centered uniform distributions; and for the family $\{f_\theta(x) = \frac{\theta}{2} e^{-\theta|x|} : \theta > 0\}$ of double exponentials. It need not be constant or bounded, however, as is shown by the family $\{F_\theta : \theta > 0\}$, where $dF_\theta(\theta) = dF_\theta(-\theta) = \frac{1}{2\theta}$ and $dF_\theta(0) = 1 - 1/\theta$, where it turns out that $(\gamma_\theta/\sigma_\theta)^3 = \sqrt{\theta}$. Whether (2.2.1) converges in this latter case will depend on the mixing measure μ.

2.2 CENTRAL LIMIT THEOREMS USING MARTINGALE METHODS

Arrays with exchangeable rows, in both the cases of finite and infinite row length, are treated using a martingale approach [Weber (1980)]. Then by invoking the conditional independence of exchangeable random variables - a property not utilized by the martingale method of proof but a fundamental property of infinite exchangeable sequences - we obtain as corollaries results for arrays with infinite exchangeable rows which were originally obtained by using the de Finetti representation [(1.3.3) and Corollary 1.2.3]. We compare these with results obtained by Weber (1980) for the special weights $a_{nk} = n^{-1/2}$, $k = 1, \ldots, n$, using reverse martingale techniques (which do not seem to work for more general weights). We present also the result of Chernoff and Teicher (1958) and Weber (1980) for a triangular array of exchangeable random variables with constant row-sums.

We first list a central limit theorem for martingale difference arrays which follows from Theorem 3.2 of Hall and Heyde (1980), which will be needed in what follows.

An array $\{X_{nk}, \mathcal{F}_{nk} : 1 \leq k \leq k_n, n \geq 1\}$ of random variables and σ-fields is a martingale difference array if for each n, X_{nk} is $\mathcal{F}_{nk}$- measurable, $\mathcal{F}_{nk} \subset \mathcal{F}_{n,k+1}$, and $E(X_{nk}|\mathcal{F}_{n,k-1}) = 0$ a.s., for $k = 1, \ldots, k_n$.

Theorem 2.2.1 Let $\{X_{nk}, \mathcal{F}_{nk} : 1 \leq k \leq k_n, n \geq 1\}$ be a martingale difference array and assume:

(i) $\max_{1 \le k \le k_n} |X_{nk}| \xrightarrow{P} 0$;

(ii) $\sup_n E \max_{1 \le k \le k_n} X_{nk}^2\} < \infty$;

(iii) $\Sigma_{k=1}^{k_n} X_{nk}^2 \xrightarrow{P} \eta^2$, a random variable, a.s. finite, and F_0- measurable, where F_0 is the completion of $\bigcap_{n=1}^{\infty} F_{n0}$.

Then $S_n = \Sigma_{k=1}^{k_n} X_{nk} \xrightarrow{d} Z$, a random variable, mixture of normals, with characteristic function $E[e^{-\frac{1}{2} t^2 \eta^2}]$.

<u>Proof</u>: First, suppose that η^2 is a.s. bounded. Thus, for some $C > 1$

$$P(\eta^2 < C) = 1. \tag{2.2.3}$$

Let $X'_{ni} = X_{ni} I(\Sigma_{j=1}^{i-1} X_{nj}^2 \le 2C)$ and $S'_{ni} = \Sigma_{j=1}^{i} X'_{nj}$. Then $\{S'_{ni}, F_{ni}\}$ is a martingale array. Since

$$P(X'_{ni} \ne X_{ni} \text{ for some } i \le k_n) \le P(U_{nk_n}^2 > 2C) \to 0 \tag{2.2.4}$$

we have $P(S'_{nk_n} \ne S_{nk_n}) \to 0$, and so

$$E|\exp(itS'_{nk_n}) - \exp(itS_{nk_n})| \to 0.$$

Hence $S_{nk_n} \xrightarrow{d} Z$ (stably) if and only if $S'_{nk_n} \xrightarrow{d} Z$ (stably). By Problem 2.5 if suffices to show that $T'_n(t) = \prod_{j=1}^{k_n} (1 + itX'_{nj}) \to 1$ weakly in L^1. Let

$$J_n = \begin{cases} \min\{i \le k_n | U_{ni}^2 > 2C\} & \text{if } U_{nk_n}^2 > 2C \\ k_n & \text{otherwise.} \end{cases}$$

Then

$$E|T'_n|^2 = E\prod_j(1+t^2X'^2_{nj}) \leq E[\{\exp(t^2\sum_{j=1}^{J_n-1}X'^2_{nj})\}(1+t^2X^2_{nJ_n})]$$

$$\leq \{\exp(2Ct^2)\}(1+t^2EX^2_{nJ_n}),$$

which is bounded uniformly in n, by (2.2.2). Consequently $\{T'_n\}$ is uniformly integrable.

Let $m \geq 1$ be fixed and let $E \in F_{mk_m}$; then $E \in F_{nk_n}$ for all $n \geq m$. For such an n,

$$E[T'_nI(E)] = E[I(E)\prod_{j=1}^{k_n}(1 + itX'_{nj})]$$

$$= E[I(E)\prod_{j=1}^{k_m}(1+itX'_{nj})\prod_{j=k_m+1}^{k_n}E(1+itX'_{nj}|F_{n,j-1})]$$

$$= E[I(E)\prod_{j=1}^{k_m}(1 + itX'_{nj})] = P(E) + R_n,$$

where the remainder term R_n consists of at most $2^{k_m} - 1$ terms of the form

$$E[I(E)(it)^rX'_{nj_1}\ldots X'_{nj_r}],$$

where $1 \leq r \leq k_m$ and $1 \leq j_1 \leq j_2 \leq \ldots \leq j_r \leq k_m$. Since

$$|X'_{nj1}\ldots X'_{njr}|^2 \leq (\sum_{j=1}^{J_n-1}X'^2_{nj})^{r-1}(\max_i X'^2_{ni})$$

$$\leq (2C)^{r-1}(\max_i X'^2_{ni}),$$

it follows that

$$|R_n| \leq (2^{k_m}-1)(2C)^{k_m/2}E(\max_i|X'_{ni}|).$$

But, for any $\varepsilon > 0$,

$$E(\max_i |X_{ni}| \le \varepsilon + E[\max_i |X_{ni}| I(|X_{ni}| > \varepsilon)]$$

$$= \varepsilon + E[(\max_i |X_{ni}|) I(\max_i |X_{ni}| > \varepsilon)]$$

$$\le \varepsilon + [E(\max_i X_{ni}^2) P(\max_i |X_{ni}| > \varepsilon)]^{\frac{1}{2}}$$

$$\to \varepsilon \quad \text{as} \quad n \to \infty.$$

It follows that $E(\max_i |X_{ni}|) \to 0$ and so $R_n \to 0$. Therefore,

$$E[T_n' I(E)] \to P(E). \tag{2.2.5}$$

Let $\mathcal{F}_\infty$ be the σ-field generated by $\bigcup_1^\infty \mathcal{F}_{no}$. For any $E' \in \mathcal{F}_\infty$ and any $\varepsilon > 0$ there exists an m and an $E \in \mathcal{F}_{mk_m}$ such that $P(E \,\Delta\, E') < \varepsilon$. Since $\{T_n'\}$ is uniformly integrable and

$$|E[T_n' I(E')] - E[T_n' I(E)]| \le E[|T_n'| I(E \,\Delta\, E')],$$

$\sup_n |E[T_n' I(E')] - E[T_n' I(E)]|$ can be made arbitrarily small by choosing ε sufficiently small. It now follows from (2.2.5) that for any $E' \in \mathcal{F}_\infty$, $E[T_n' I(E')] \to P(E')$. This in turn implies that for any bounded $\mathcal{F}_\infty$-measurable r.v. X, $E[T_n' X] \to E(X)$. Finally, if $E \in \mathcal{A}$, then

$$E[T_n' I(E)] = E[T_n' E(I(E)|\mathcal{F}_\infty)] \to E[E(I(E)|\mathcal{F}_\infty)] = P(E).$$

This establishes $T_n' \xrightarrow{d} 1$ and completes the proof in the special case where (2.2.3) holds.

It remains only to remove the boundedness condition (2.2.3). If η^2 is not a.s. bounded, then given $\varepsilon > 0$, choose a continuity point C of η^2 such that $P(\eta^2 > C) > \varepsilon$. Let

$$\eta_c^2 = \eta^2 I(\eta^2 \le C) + CI(\eta^2 > C),$$

$$X''_{ni} = X_{ni} I\Big(\sum_{j=1}^{i-1} X^2_{nj} \le C\Big) \quad \text{and} \quad S''_{ni} = \sum_{j=1}^{i} X''_{nj}.$$

Then $\{S''_{ni}, F_{ni}\}$ is a martingale array, and conditions (2.2.1) and (2.2.2) are satisfied. Now,

$$\Big(\sum_i X^2_{ni}\Big) I\Big(\sum_i X^2_{ni} \le C\Big) + C I\Big(\sum_i X^2_{ni} > C\Big) \le \sum_i X''^2_{ni}$$

$$\le \Big(\sum_i X^2_{ni}\Big) I\Big(\sum_i X^2_{ni} \le C\Big) + \Big(C + \max_i X^2_{ni}\Big) I\Big(\sum_i X^2_{ni} > C\Big).$$

Since C is a continuity point of the distribution of η^2,

$$I\Big(\sum_i X^2_{ni} \le C\Big) \xrightarrow{P} I(\eta^2 \le C),$$

and so

$$\sum_i X''^2_{ni} \xrightarrow{P} \eta^2_c.$$

As η^2_c is a.s. bounded, the first part of the proof tells us that $S''_n \xrightarrow{d} Z_c$, where the r.v. Z_c has characteristic function $E\exp(-\tfrac{1}{2}\eta^2_c t^2)$. If $E \in A$, then

$$|E[I(E)\exp(itS_{nk_n})] - E[I(E)\exp(-\tfrac{1}{2}\eta^2 t^2)]|$$

$$\le E|\exp(itS_{nk_n}) - \exp(itS''_{nk_n})| + |E[I(E)\exp(itS''_{nk_n})]$$

$$-E[I(E)\exp(-\tfrac{1}{2}\eta^2_c t^2)]| + E|\exp(-\tfrac{1}{2}\eta^2_c t^2) - \exp(-\tfrac{1}{2}\eta^2 t^2)|.$$

Since $S''_{nk_n} \to Z_c$, the second term on the right-hand side converges to zero as $n \to \infty$. The first and third terms are each less than 2ε in the limit since

$$P(S_{nk_n} \ne S''_{nk_n}) \le P(X''_{ni} \text{ for some } i)$$

$$\le P(U^2_{nk_n} > C) \to P(\eta^2 > C) < \varepsilon.$$

Hence for all $\varepsilon > 0$,

$$\limsup_{n\to\infty} |E[I(E)\exp(itS_{nk_n})] - E[I(E)\exp(-\tfrac{1}{2}\eta^2 t^2)]| \leq 4\varepsilon.$$

It follows that the limit equals zero, and with Problem 2.5 completes the proof.

///

We first consider the case of an array with infinite rows and begin by constructing certain arrays of σ-fields.

Let $\{X_{nk} : k, n \geq 1\}$ be an array of row-wise exchangeable random variables with $E|X_{n1}| < \infty$, for all $n \in N$. Let

$$F^m_{nk} = \sigma\{X_{n1}, \ldots, X_{nk}, \sum_{j=k+1}^{m} X_{nj}, X_{n,m+1}, X_{n,m+2}, \ldots\},$$

for $k = 0, 1, \ldots, m - 1$; and $G_n = \bigcap_{m=1}^{\infty} F^m_{n0}$.

Put $M_{nm} = m^{-1} \Sigma^m_{j=1} X_{nj}$ and note that $\{M_{nm}, F^m_{n0} : m \geq 1\}$ is a reverse martingale, for each $n \in N$. Since $E|M_{nm}| \leq E|X_{n1}| < \infty$, $M_{nm} \xrightarrow{L1} E(X_{n1}|G_n)$, as $m \to \infty$, for each $n \in N$.

Choose a sequence $\{\varepsilon_n\}$ of positive numbers, $\varepsilon_n \to 0$. Consider a sequence $\{k_n\}$, $k_n \in N$, $k_n \uparrow \infty$; and choose a second sequence $\{m_n\}$, $m_n \in N$, $m_n \geq k_n$, $m_n \uparrow \infty$, such that $\lim_{n\to\infty} k_n m_n^{-1} = 0$ and such that

$E|E(X_{nk}|F^{m_n}_{n,k-1}) - E(X_{n1}|G_n)| < \varepsilon_n$, for every $k = 1, \ldots, k_n$ and $n \in N$. This is possible since

$$E|E(X_{nk}|F^m_{n,k-1}) - E(X_{n1}|G_n)| =$$

$$= E\left|\frac{1}{m-k+1} \Sigma_{j=k}^{m} X_{nj} - E(X_{n1}|G_n)\right|$$

$$= E\left|\frac{m}{m-k+1} M_{nm} - \frac{1}{m-k+1} \Sigma_{j=1}^{k-1} X_{nj} - E(X_{n1}|G_n)\right|$$

$$\le \frac{k_n-1}{m-k_n+1} E|M_{nm}| + \frac{k_n-1}{m-k_n+1} E|X_{n1}| +$$

$$+ E|M_{nm} - E(X_{n1}|G_n)| \to 0,$$

as $m \to \infty$, for each $n \in N$.

Having chosen the sequence $\{m_n\}$, we henceforth write simply F_{nk} for $F_{nk}^{m_n}$, $k = 0,1, \ldots, m_n - 1$. Define the further σ-field F_0 to be the completion of $\bigcap_{n=1}^{\infty} F_{n0}$.

The almost sure equality

$$E(X_{nk}|F_{n,k-1}) = \frac{\Sigma_{j=k}^{m_n} X_{nj}}{m_n-k+1}, \text{ a.s.,}$$

$k = 0,1, \ldots, m_n$, which follows from Property 1.2.4, is instrumental in the proofs that follow.

Theorem 2.2.2 Let $\{X_{nk} : n,k \ge 1\}$ be an array of row-wise exchangeable random variables with $\sup_n E X_{n1}^2 < \infty$. Given $\{\varepsilon_n\}$, $\varepsilon_n > 0$, $\varepsilon_n \to 0$, let the sequences $\{k_n\}$ and $\{m_n\}$ and the σ-fields $\{F_{nk}\}$, $k = 0,., \ldots, m_n - 1$, be chosen as above. Let $\{a_{nk} : 1 \le k \le k_n, n \ge 1\}$ be an array of real constants and set $\bar{a}_n = \max_{1\le k\le k_n} |a_{nk}|$ and $A_n = \Sigma_{k=1}^{k_n} a_{nk}^2$.

Assume:

1° $\varepsilon_n \Sigma_{k=1}^{k_n} |a_{nk}| \to 0$, as $n \to \infty$;

2° $\sup_n A_n < \infty$;

3° $EX_{n1}X_{n2} = o(A_n^{-1})$;

4° $\max_{1\leq k\leq m_n} |X_{nk}| = o_p(\bar{a}_n^{-1})$;

5° $\Sigma_{k=1}^{k_n} a_{nk}^2 X_{nk}^2 \xrightarrow{P} \eta^2$, an F_0-measurable random variable, a.s. finite

Then $\Sigma_{k=1}^{k_n} a_{nk}(X_{nk} - E(X_{n1}|G_n)) \xrightarrow{d} Z$, a mixture of normals, with characteristic function $Ee^{-\frac{1}{2}t^2\eta^2}$.

Proof: Put $Y_{nk} = a_{nk}(X_{nk} - E(X_{nk}|F_{n,k-1}))$. Then $\{Y_{nk} : 1 \leq k \leq k_n, n \geq 1\}$ is a martingale difference array, for which we verify the conditions of Theorem 2.2.1. An indicated "k", on $\max_k$, Σ_k, etc., will always be taken from 1 to k_n.

For (i) of Theorem 2.2.1 we have

$$\max_k |Y_{nk}| \leq \max_k |a_{nk}||X_{nk}| + \max_k |a_{nk}||E(X_{nk}|F_{n,k-1})|$$

$$\leq \max_k |a_{nk}|[\max_k|X_{nk}| + \max_k |\frac{\Sigma_{j=k}^{m_n} X_{nj}}{m_n-k+1}|]$$

$$\leq 2\bar{a}_n \max_{1\leq k\leq m_n} |X_{nk}| \xrightarrow{P} 0, \text{ by } 4°.$$

For (ii) we have

$$E\{\max_k Y_{nk}^2\} \leq 2E\{\max_k a_{nk}^2 X_{nk}^2 + \max_k a_{nk}^2 E^2(X_{nk}|F_{n,k-1})\}$$

$$\leq 2E\{\Sigma_k a_{nk}^2 X_{nk}^2 + \Sigma_k a_{nk}^2 E(X_{nk}^2|F_{n,k-1})\}$$

$$\leq 4 \sup_n EX_{n1}^2 \sup_n A_n < \infty, \text{ by } 2°.$$

To get (iii) write

$$\Sigma_k Y_{nk}^2 = \Sigma_k a_{nk}^2 X_{nk}^2 + \Sigma_k a_{nk}^2 E^2(X_{nk}|F_{n,k-1})$$
$$- 2\,\Sigma_k a_{nk}^2 X_{nk} E(X_{nk}|F_{n,k-1}). \tag{2.2.6}$$

Now $E\,\Sigma_k a_{nk}^2 E^2(X_{nk}|F_{n,k-1})$

$$= |\Sigma_k a_{nk}^2 E(X_{nk} E(X_{nk}|F_{n,k-1}))|$$

$$\leq \Sigma_k a_{nk}^2 |EX_{nk} \frac{\Sigma_{j=k}^{m_n} X_{nj}}{m_n-k+1}| \tag{2.2.7}$$

$$= EX_{n1}^2\, \Sigma_k \frac{a_{nk}^2}{m_n-k+1} + |EX_{n1}X_{n2}|\, \Sigma_k \frac{m_n-k}{m_n-k+1}\, a_{nk}^2 \leq$$

$$\leq \sup_n EX_{n1}^2 \frac{\sup A_n}{m_n-k_n+1} + A_n|EX_{n1}X_{n2}| \to 0,$$

by 2°, $k_n m_n^{-1} \to 0$ and 3°. Hence 5° and (2.2.6) yield $\underset{n\to\infty}{\text{p-lim}}\, \Sigma_k Y_{nk}^2 = \underset{n\to\infty}{\text{p-lim}}\, \Sigma_k a_{nk}^2 X_{nk}^2 = \eta^2$, which is (iii) of Theorem 2.2.1.

We thus have

$$\Sigma_k Y_{nk} = \Sigma_k a_{nk}(X_{nk} - E(X_{nk}|F_{n,k-1})) \overset{d}{\to} Z, \tag{2.2.8}$$

a mixture of normals, with characteristic function $Ee^{-\frac{1}{2}t^2\eta^2}$.

Finally, noting that

$$E|\Sigma_k a_{nk}(E(X_{nk}|F_{n,k-1}) - E(X_{n1}|G_n))|$$

$$\leq \Sigma_k |a_{nk}|\, E|\, E(X_{nk}|F_{n,k-1}) - E(X_{n1}|G_n)|$$

$$< \varepsilon_n \Sigma_k |a_{nk}| \to 0, \text{ by } 1°, \text{ as } n \to \infty,$$

we get

$$\Sigma_k a_{nk}(E(X_{nk}|F_{n,k-1}) - E(X_{n1}|G_n)) \xrightarrow{P} 0,$$

and finally, from (2.2.7),

$$\Sigma_k a_{nk}(X_{nk} - E(X_{n1}|G_n)) \xrightarrow{d} Z. \quad ///$$

<u>Remark 2.2.1</u> If $\underline{\lim}_{n\to\infty} A_n = 0$, then convergence in Theorem 2.2.2 is degenerate, to the probability distribution concentrated at zero. Indeed, if $\{n'\}$ is a subsequence of the natural numbers such that $A_{n'} \to 0$, then

$$P[\Sigma_{k=1}^{k_{n'}} a_{n'k}^2 X_{n'k}^2 > \varepsilon] < \varepsilon^{-1} (\sup_n EX_{n1}^2) A_{n'} \to 0, \text{ for all } \varepsilon > 0.$$

This, together with 5° of Theorem 2.2.2, yields $\eta^2 = 0$ a.s.

<u>Remark 2.2.2</u> Note that one can allow for any rate of growth of $\Sigma_{k=1}^{k_n} |a_{nk}|$ by appropriate choice of the sequence $\{\varepsilon_n\}$ (and consequently $\{m_n\}$). However, the rate of decrease of $\bar{a}_n$ is still determined by the rate of growth (or decrease) of $\max_{1 \le k \le m_n} |X_{nk}|$, that is, by the properties of the array $\{X_n\}$.

If we now introduce the property of conditional independence of an infinite exchangeable sequence of random variables, the centerings of the random variables of the array $\{X_{nk}\}$ at conditional expectations can be eliminated. Since the σ-field G_n contains the tail σ-field and is contained in the exchangeable σ-field of the nth row $\{X_{nk}: n, k \ge 1\}$ of the array, the exchangeable sequence $\{X_{nk}: n, k \ge 1\}$ is conditionally independent, given G_n [rf. Chow & Teicher (1978), Section 7.3 or Olshen (1974)]. This will be used in the proof of the following result.

<u>Corollary 2.2.3</u> If in addition to the hypotheses of Theorem 2.2.2 the following holds:

$$EX_{n1}X_{n2} = o((\Sigma_{k=1}^{k_n} a_{nk})^{-2}),$$

then $\Sigma_{k=1}^{k_n} a_{nk}X_{nk} \xrightarrow{d} Z,$

with Z as in Theorem 2.2.1.

<u>Proof</u>: Indeed, $\Sigma_{k=1}^{k_n} a_{nk}E(X_{n1}|G_n) \xrightarrow{P} 0$, since $E(\Sigma_{k=1}^{k_n} a_{nk}E(X_{n1}|G_n))^2 = (\Sigma_{k=1}^{k_n} a_{nk})^2 E(E^2(X_{n1}|G_n))$ $= (\Sigma_{k=1}^{k_n} a_{nk})^2 E(E(X_{n1}X_{n2}|G_n)) = (\Sigma_{k=1}^{k_n} a_{nk})^2 EX_{n1}X_{n2} \to 0$. ///

For the weights $a_{nk} = n^{-1/2}$, $k = 1, \ldots, n$, the condition of Corollary 2.2.3 becomes $EX_{n1}X_{n2} = o(\frac{1}{n})$, the condition appearing in Theorem 2 of Blum et al. (1958) and in Corollaries 1 & 2 of Weber (1980).

We state, for comparison and reference, Corollary 2 of Weber (1980), which was obtained by using reverse martingales.

<u>Theorem 2.2.4</u> (Weber) Let $\{X_{nk}: n,k \geq 1\}$ be a row-wise exchangeable array, with $EX_{n1}^4 < \infty$, $\forall n \in N$. Assume

(a) $EX_{n1}X_{n2} = o(n^{-1})$;

(b) $EX_{n1}^2X_{n2}^2 \to 1$;

(c) $E(X_{n1}^2 I_{[|X_{n1}| > \varepsilon\sqrt{n}]}) \to 0, \quad \forall \varepsilon > 0.$

Then $\frac{1}{\sqrt{n}} \Sigma_{k=1}^{n} X_{nk} \xrightarrow{d} N(0,1)$.

This result is slightly stronger than Theorem 2 of Blum et al. (1958). Note that for a single exchangeable sequence $X_1, X_2, \ldots$, Theorem 2.2.4 reduces to the direct part of Corollary 2.1.3.

We turn next to triangular arrays of finite row length, with exchangeable rows.

For a triangular array $\{X_{nk}\}$, $k = 1, \ldots, m_n$, of random variables, define the σ-fields $\{\mathcal{F}_{nk}\}$ by

$$\mathcal{F}_{nk} = \sigma\{X_{n1}, \ldots, X_{nk}, \Sigma_{j=k+1}^{m_n} X_{nj}\},\ k = 0, 1, \ldots, m_n - 1.$$

Denote by $\mathcal{F}$ the completion of $\bigcap_{n=1}^{\infty} \mathcal{F}_{n0}$.

Theorem 2.2.5 Let $\{X_{nk}\}$, $k = 1, \ldots, m_n$, $m_n \to \infty$, be an array of row-wise exchangeable random variables with $\sup_n EX_{n1}^2 < \infty$. Let the σ-fields $\{\mathcal{F}_{nk}\}$ be defined as above. Let $\{k_n\}$, $k_n \in N$, be a sequence such that $k_n \leq m_n$, $k_n \uparrow \infty$, and $k_n m_n^{-1} \to 0$. Let $\{a_{nk}\}$, $k = 1, \ldots, k_n$, be an array of real constants and put $\bar{a}_n = \max_{1 \leq k \leq k_n} |a_{nk}|$ and $A_n = \Sigma_{k=1}^{k_n} a_{nk}^2$. Assume

1° $\sup_n A_n < \infty$;

2° $\bar{a}_n = o(m_n^{1/2} k_n^{-1})$;

3° $EX_{n1}X_{n2} = o(\bar{a}_n^{-2} k_n^{-2})$;

4° $\max_{1 \leq k \leq m_n} |X_{nk}| = o_p(\bar{a}_n^{-1})$;

5° $\Sigma_{k=1}^{k_n} a_{nk}^2 X_{nk}^2 \xrightarrow{P} \eta^2$, an $\mathcal{F}$-measurable random variable, a.s. finite.

Then $S_n = \Sigma_{k=1}^{k_n} a_{nk}X_{nk} \overset{d}{\to} Z$, a mixture of normals with ch.f. $Ee^{-\frac{1}{2}t^2\eta^2}$.

<u>Proof</u>: Put $Y_{nk} = a_{nk}(X_{nk} - E(X_{nk}|F_{n,k-1}))$. Then $\{Y_{nk}, F_{nk} : 1 \le k \le m_n, n \ge 1\}$ is a martingale difference array for which we verify the hypotheses of Theorem 2.2.1.

(i) and (ii) of Theorem 2.2.1 are obtained exactly as in the proof of Theorem 2.2.2, using hypotheses 4° and 1°, respectively. (iii) is obtained almost exactly as in Theorem 2.2.2. The only difference is that in (2.2.7) we write

$$|EX_1X_2| \Sigma_k \frac{m_n - k}{m_n - k + 1} a_{nk}^2 \le |EX_1X_2| \Sigma_k a_{nk}^2 \le |EX_1X_2| k_n \bar{a}_n^2 =$$

$|EX_1X_2| \frac{k_n^2 \bar{a}_n^2}{k_n}$, which tends to zero by 3°. Thus 5° yields

$$\text{p-lim}_{n\to\infty} \Sigma_k Y_{nk}^2 = \text{p-lim}_{n\to\infty} \Sigma_k a_{nk}^2 X_{nk}^2 = \eta^2,$$

and it follows from Theorem 2.2.1 that

$$\Sigma_k Y_{nk} = \Sigma_k a_{nk}(X_{nk} - E(X_{nk}|F_{n,k-1})) \overset{d}{\to} Z,$$

with Z as in the conclusion of the theorem.

Finally, we show that

$$\Sigma_{k=1}^{k_n} a_{nk} E(X_{nk}|F_{n,k-1}) \overset{P}{\to} 0. \qquad (2.2.9)$$

We show L^2 convergence:

$$E(\Sigma_{k=1}^{k_n} a_{nk} E(X_{nk}|F_{n,k-1}))^2$$

$$= \Sigma_{k=1}^{k_n} a_{nk}^2 E^2(X_{nk}|F_{n,k-1})$$

$$+ 2\, E\, \Sigma_{i=1}^{k_n-1} \Sigma_{j=i+1}^{k_n} a_{ni} a_{nj} E(X_{ni} | F_{n,i-1}) E(X_{nj} | F_{n,j-1}).$$

It was shown above that the first term converges to zero. For the second term we have

$$|2\, \Sigma_{i=1}^{k_n-1} \Sigma_{j=i+1}^{k_n} a_{ni} a_{nj} E(E(X_{nj} E(X_{ni} | F_{n,i-1}) | F_{n,j-1}))|$$

$$\leq 2\, \Sigma_{i=1}^{k_n-1} \Sigma_{j=i+1}^{k_n} |a_{ni}||a_{nj}|\, |E(X_{nj} E(X_{ni} | F_{n,i-1}))|$$

$$= 2\, \Sigma_{i=1}^{k_n-1} \Sigma_{j=i+1}^{k_n} |a_{ni}||a_{nj}|\, |E\{X_{nj} \frac{\Sigma_{\ell=i}^{m_n} X_{n\ell}}{m_n-i+1}\}|$$

$$= 2\, EX_{n1}^2\, \Sigma_{i=1}^{k_n-1} \Sigma_{j=i+1}^{k_n} \frac{|a_{ni}||a_{nj}|}{m_n-i+1}$$

$$+ 2\, |EX_{n1}X_{n2}|\, \Sigma_{i=1}^{k_n-1} \Sigma_{j=i+1}^{k_n} \frac{m_n-i}{m_n-i+1} |a_{ni}||a_{nj}|$$

$$\leq 2\, \bar{a}_n^2 \sup_n EX_{n1}^2\, \Sigma_{i=1}^{k_n-1} \frac{k_n-i}{m_n-i+1}$$

$$+ 2\, \bar{a}_n^2\, |EX_{n1}X_{n2}|\, \Sigma_{i=1}^{k_n-1} (k_n-i) = I + II.$$

Since

$$\frac{k_n}{m_n} > \frac{k_n-i}{m_n-i+1} \quad \text{for } i = 1, \ldots, k_n - 1,$$

$\bar{a}_n^2\, \Sigma_{i=1}^{k_n-1} \frac{k_n-i}{m_n-i+1} < \bar{a}_n^2 \frac{k_n^2}{m_n} \to 0$ by 2°, hence $I \to 0$. Also,

$$II \leq 2\, \bar{a}_n^2\, |EX_{n1}X_{n2}|\, \frac{k_n^2 + k_n}{2} < 2\, |EX_{n1}X_{n2}|\bar{a}_n^2 k_n^2 \to 0, \text{ by } 3°.$$

Thus (2.2.8) holds and we finally have

$$\Sigma_{k=1}^{k_n} a_{nk}X_{nk} \xrightarrow{d} Z. \qquad ///$$

Remark 2.2.3 Suppose, as will normally be the case, that $\bar{a}_n \to 0$. This, together with $\sup_n A_n < \infty$, implies that $\Sigma_{k=1}^{k_n} \frac{a_{nk}^2}{k_n-k+1} \to 0$. (See problem 2.4)

(2.2.4) then shows that (iii) of Theorem 2.2.1 can be obtained by taking $k_n = m_n$. Thus (striking hypothesis 2°) one can still conclude that $\Sigma_{k=1}^{k_n} a_{nk}(X_{nk} - E(X_{nk}|F_{n,k-1})) \xrightarrow{d} Z$. In practice one would like to be rid of the awkward centering, and for this one needs both sequences $\{k_n\}$ and $\{m_n\}$, with 2° and $k_n m_n^{-1} \to 0$ satisfied.

Remark 2.2.4 Hypotheses 2° and 3° of Theorem 2.2.5 are stronger than needed. It follows from the proof that $\Sigma_{i=1}^{k_n-1} \Sigma_{j=i+1}^{k_n} \frac{|a_{ni}a_{nj}|}{m_n-i+1} \to 0$ and

$EX_{n1}X_{n2} = O([\Sigma_{i=1}^{k_n-1} \Sigma_{j=i+1}^{k_n}|a_{ni}a_{nj}|]^{-1})$ are sufficient for the conclusion of the theorem. Note also that if 2° is satisfied, then 3° will also be satisfied whenever $EX_{n1}X_{n2} = O(m_n^{-1})$.

Taking $a_{nk} = n^{-1/2}$ for $k = 1, \ldots, n$, we have the following result.

Corollary 2.2.6 Let $\{X_{nk}\}$ $1 \le k \le n \le m_n$, with $nm_n^{-1} \to 0$, be an array of row-wise exchangeable random variables and let the σ-fields $\{F_{nk}\}$, $k = 1, \ldots, m_n$, be defined as in Theorem 2.2.4. Assume

(a) $\sup_n EX_{n1}^2 < \infty$;

(b) $EX_{n1}X_{n2} = o(n^{-1})$;

(c) $\max_{1\le k\le m_n} |X_{nk}| = o_p(\sqrt{n})$;

(d) $n^{-1}\Sigma_{k=1}^{n} X_{nk}^2 \xrightarrow{P} \eta^2$, an F_o-measurable random variable, a.s. finite.

Then $n^{-1/2}\Sigma_{k=1}^{n} X_{nk} \xrightarrow{d} Z$, a mixture of normals, with ch.f. $Ee^{-\frac{1}{2}t^2\eta^2}$.

Corollary 2.2.6 (Except for convergence to Brownian motion) is a generalization of Corollary 1 of Weber (1980).

2.3 PROBLEMS

2.1 Argue that if $\{a_{nk}\}$ are non-negative weights satisfying (a) and (b) of Theorem 2.1.2, and if $\{X_k\}$ is an exchangeable sequence with $EX_1 = 0$, $EX_1^2 < \infty$, that if $EX_1X_2 > 0$, then $\Sigma_{k=1}^{k_n} a_{nk}X_k$ cannot converge in distribution (Lemma 2.1.1 with $r = 1$ is useful here).

2.2 Find a counterexample to the converse of Theorem 2.1.2 which is described in Remark 2.1.1.

2.3 Prove that if $\int_R \Phi(\sigma^{-1}x)dG(\sigma) = \Phi(x)$, $\forall\, x \in R$, where G is a distribution function concentrated on $[0,\infty)$ and Φ is the normal cumulative, that then $G(\sigma) = I_{[1,\infty)}(\sigma)$.

2.4 Prove that $\max_{1\le k\le k_n} |a_{nk}| \to 0$ and $\sup_n \Sigma_{k=1}^{k_n} a_{nk}^2 < \infty$ imply $\Sigma_{k=1}^{k_n} \frac{a_{nk}^2}{k_n-k+1} \to 0$.

2.5 In Theorem 2.2.1 show that $T_n(t) = \Pi_{j=1}^{k_n}(1+itX_{nj}) \to 1$ weakly in L^1 implies $S_n \xrightarrow{d} Z$. (Lemma 3.1 of Hall and Heyde (1980).)

CHAPTER III

STOCHASTIC CONVERGENCE OF WEIGHTED SUMS OF EXCHANGEABLE RANDOM VARIABLES

3.0 INTRODUCTION

This chapter will contain some of the standard laws of large numbers and some of weighted sum convergence results for random variables. Section 3.1 will list some weak convergence results for independent and exchangeable random variables. In Section 3.2, strong convergence results for weighted sums of independent and exchangeable random variables will be presented. Finally, in Section 3.3, strong convergence results for triangular arrays of exchangeable random variables will be presented. An example illustrating the results of Section 3.3 will also be presented.

3.1 WEAK CONVERGENCE OF WEIGHTED SUMS OF RANDOM VARIABLES

Of considerable interest in this section is the convergence of the weighted sum, $S_n = \Sigma_{k=1}^{n} a_{nk} X_{nk}$, under the various conditions of the weights $\{a_{nk}\}$ and random variables $\{X_{nk}\}$. Toeplitz sequences are examples of a certain class of weights. Certain conditions on Toeplitz sequences are needed to produce convergence results for such weighted sums. These results will be used primarily as reference and comparisons in the later presentation of results for separable Banach spaces. The reader can find the proofs in the various referenced journals where the results have appeared.

Theorem 3.1.1 (Pruitt, 1966) Let $\{X_k\}$ be independent,

identically distributed non-degenerate random variables such that $E|X_1| < \infty$ and let $\{a_{nk}\}$ be a Toeplitz sequence, such that $\lim_{n\to\infty} \Sigma_{k=1}^{\infty} a_{nk} = 1$. A necessary and sufficient condition that $Y_n = \Sigma_{k=1}^{\infty} a_{nk} X_k \to EX_1$ in probability is that $\max_k |a_{nk}| \to 0$

The following result by Rohatgi (1971), dealing with independent but not necessarily identically distributed random variables $\{X_k\}$, requires uniformly bounded tail probabilities, that is, the existence of a random variable X such that

$$P[|X_n| \geq t] \leq P[|X| \geq t] \quad (3.1.1)$$

for all $t > 0$ and all n.

<u>Theorem 3.1.2</u> Let $\{X_n\}$ be a sequence of independent random variables such that Condition (3.1.1) holds for some random variable X and let $\{a_{nk}\}$ be a Toeplitz sequence. If

(i) $\max_k |a_{nk}| \to 0$ as $n \to \infty$

(ii) $E|X|^r < \infty$ for some $0 < r \leq 1$, and

(iii) $\Sigma_{k=1}^{\infty} |a_{nk}|^r \leq C$ for each n and some constant $C > 0$, then

$Y_n = \Sigma_{k=1}^{\infty} a_{nk} X_k \to 0$ in probability. (If $r = 1$, then also assume that $EX_n = 0$ for all n.)

The next result by Taylor (1982) considers arrays of random variables. A condition imposed on the weights $\{a_{nk}\}$ and the measure of nonorthogality condition assumed on the array of random variables $\{X_{nk}\}n,k$ are used to insure convergence in probability.

Theorem 3.1.3 (Taylor, 1982) Let $\{X_{nk}\}n,k$ be real-valued random variables such that $E(X_{nk}^2) \leq \Gamma < \infty$ for all n and k and let $\rho_n = \sup_{k \neq \ell} |E(X_{nk}X_{n\ell})|$ be the measure of nonorthogonality. If $\{a_{nk}\}$ are constants such that

$$\Sigma_{k=1}^n a_{nk}^2 \to 0 \text{ and} \tag{3.1.2}$$

$$\rho_n(\Sigma_{k=1}^n |a_{nk}|)^2 \to 0, \text{ then} \tag{3.1.3}$$

$$\Sigma_{k=1}^n a_{nk}X_{nk} \to 0 \text{ in probability.}$$

Proof: By Markov's inequality, for $\varepsilon > 0$,

$$P[|\Sigma_{k=1}^n a_{nk}X_{nk}| > \varepsilon] \leq \frac{1}{\varepsilon^2} E(\Sigma_{k=1}^n a_{nk}X_{nk})^2$$

$$\leq \frac{1}{\varepsilon^2}(E(\Sigma_{k=1}^n a_{nk}^2 X^2{}_{nk} + \Sigma\Sigma_{\ell \neq k} |a_{nk}a_{n\ell}||X_{nk}X_{n\ell}|))$$

$$= \frac{1}{\varepsilon^2}(\Sigma_{k=1}^n a_{nk}^2 E(X_{nk}^2) + \Sigma\Sigma_{\ell \neq k} |a_{nk}a_{n\ell}|E|X_{nk}X_{n\ell}|)$$

$$\leq \frac{1}{\varepsilon^2}(\Gamma\ \Sigma_{k=1}^n a_{nk}^2 + \Sigma\Sigma_{\ell \neq k} |a_{nk}a_n|\sup_{k \neq \ell} E|X_{nk}X_{n\ell}|)$$

$$= \frac{1}{\varepsilon^2}(\Gamma\ \Sigma_{k=1}^n a_{nk}^2 + \rho_n \Sigma\Sigma_{\ell \neq k} |a_{nk}a_{n\ell}|).$$

Now $\sum\sum_{\ell \neq k} |a_{nk}a_{n\ell}| = (\Sigma_{k=1}^{n} |a_{nk}|)^2 - (\Sigma_{k=1}^{n} a_{nk}^2)$ which implies

that

$$P[|\Sigma_{k=1}^{n} a_{nk}X_{nk}| > \varepsilon] \leq$$

$$\frac{\Gamma}{\varepsilon^2}(\Sigma_{k=1}^{n} a_{nk}^2) + \frac{\rho_n}{\varepsilon^2}[(\Sigma_{k=1}^{n}|a_{nk}|)^2 - \Sigma_{k=1}^{n} a_{nk}^2]$$

$$\rightarrow 0 \text{ as } n \rightarrow \infty. \qquad ///$$

Remark: If $\{X_{nk} : k \geq 1\}$ is exchangeable for each n, then $\rho_n = E(X_{n1}X_{n2})$, and Example 1.2.1 suggested that $\rho_n \rightarrow 0$. In Theorem 3.1.3, (3.1.3), the behavior of ρ_n is related to a growth condition on the weights.

3.2 STRONG CONVERGENCE OF WEIGHTED SUMS OF RANDOM VARIABLES

In this section, strong convergence results will be obtained for sequences, series, and infinite and triangular arrays of exchangeable random variables. Strong convergence results will be obtained for sequences of random variables that are functions of sums of exchangeable random variables. Other results in the section require a pth moment condition, $0 < p < 1$, on the exchangeable random variables. As in the previous section, conditions on weights, such as that of forming a Toeplitz array, are needed to produce strong convergence results. In the development of strong convergence results for triangular arrays of exchangeable random variables, martingale techniques will be used. Some results for sequences and infinite arrays of independent random variables will be listed for reference and comparison to results for exchangeable random variables.

The first result for i.i.d. random variables requires a moment condition on the random variables as well as a condition on the Toeplitz sequence.

A sequence $\{t_n\}$ is said to be $\mathcal{O}(f(n))$ if $\sup_n |\frac{t_n}{f(n)}| < \infty$

and is said to be $o(f(n))$ if $\lim_{n\to\infty} \left|\frac{t_n}{f(n)}\right| = 0$.

Theorem 3.2.1 (Pruitt, 1966) Let $\{X_k\}$ be independent, identically distributed random variables and let $\{a_{nk}\}$ be a Toeplitz sequence. If $\max_k |a_{nk}| = O(n^{-\gamma})$ for some $\gamma > 0$, then $E|X_1|^{1+1/\gamma} < \infty$ implies that $Y_n = \Sigma_{k=1}^{\infty} a_{nk}X_k \to EX_1$ with probability one.

The next result by Rohatgi (1971) for independent random variables requires that condition (3.11) holds as well as conditions on the Toeplitz sequence.

Theorem 3.2.2 Let $\{X_n\}$ be a sequence of independent random variables such that $EX_n = 0$ for all n and such that Condition (3.1.1) holds for some random variable X and let $\{a_{nk}\}$ be a Toeplitz sequence. If

(i) $\max_k |a_{nk}| = O(n^{-\gamma})$ for some $\gamma > 0$ and

(ii) $E|X|^{1+1/\gamma} < \infty$, then

$Y_n = \Sigma_{k=1}^{\infty} a_{nk}X_k \to 0$ with probability one.

Weights other than Toeplitz sequences were considered by Chow and Lai (1973). They considered the convergence of $Y_n = \Sigma_{k=1}^{n} a_{nk}X_k$ where $\{X_n\}$ were independent, identically distributed random variables and the array $\{a_{nk}\}$ was such that $\lim_{n\to\infty} \Sigma_{k=1}^{n} a_{nk}^2 < \infty$.

The following, Theorem 3.2.3 was also given by Chow and Lai (1973).

Theorem 3.2.3 Let $\{X_n\}$ be a sequence of independent, identically distributed random variables such that $EX_1 = 0$.

For $1 \leq \alpha \leq 2$, $E|X_1|^\alpha < \infty$ if and only if for every array $\{a_{nk}\}$ such that $\sup_n \Sigma_{k=1}^n a_{nk}^2 < \infty$, it follows that $n^{-1/\alpha} \Sigma_{k=1}^n a_{nk} X_k \to 0$ with probability one.

The next two strong convergent results are by Bru, Heinich and Lontgieter (1981). The first result is for positive sub-additive functions on R^+. Let $Y_n = \frac{1}{n}\phi(X_1+\ldots+X_n)$ where $\phi(X_1+\ldots+X_n)$ is a positive, subadditive function of positive exchangeable random variables $X_1, X_2, \ldots$. With a moment condition on $\phi(X_1)$, the following results by Bru, Heinich and Lontgieter are generalizations of the classical strong laws of large numbers. For example, $\phi(X_1+\ldots+X_n) = X_1+\ldots+X_n$ is a positive subadditive function when $X_1, X_2, \ldots$ are positive exchangeable random variables.

<u>Theorem 3.2.4</u> Let ϕ be a positive sub-additive function on R^+ such that $\phi(0) = 0$ and is continuous at the origin. Then for each sequence $\{X_n\}$ of positive exchangeable random variables such that $E(\phi(X_1)) < \infty$, $\{(\frac{1}{n})\ \phi(X_1+\ldots+X_n)\}$ is a bounded decreasing sequence in L^1 that converges almost surely.

<u>Remark</u>: For $\phi(X_1+\ldots+X_n) = X_1+\ldots+X_n$, Problem 1.9 follows immediately for positive exchangeable random variables by Theorem 3.2.4 $\frac{1}{n} \Sigma_{k=1}^n X_k \leq \frac{1}{n-1} \Sigma_{k=1}^{n-1} X_k$.

The next theorem is a strong convergence result for real-valued functions f of type(s).

<u>Definition 3.2.1</u> A function f defined on R such that $f(0) = 0$ is said to be of type(s) if $f(x)/x$ is decreasing.

Theorem 3.2.5 Let f be a function of type(s) and $\{X_n\}$ a sequence of positive exchangeable random variables with $E(f(X_1)) < \infty$. Then $(\frac{1}{n})f(X_1+\ldots+X_n)$ is a decreasing submartingale that converges almost surely in L^1.

Edgar and Sucheston (1981) have obtained strong laws of large numbers for sequences of exchangeable random variables in L_p, $0 < P < 1$.

Theorem 3.2.6 Let $\{X_n\}$ be a sequence of exchangeable random variables. If $0 < p < 1$, and $X_i \in L_p$, then $(\frac{1}{m})(X_1+\ldots+X_m)^p$ is a bounded decreasing submartingale in L_1 and $m^{-1/p}(X_1+\ldots+X_m)$ converges almost surely to zero in L_p.

Proof: Assume that $X_i \geq 0$. Let $S_m = X_1+\ldots+X_m$, $Y_{-m} = S_m^p/m$, and $F_{-m} = \sigma\{Y_{-r} : r \geq m\}$. For each permutation $\pi \in P_m$, $(X_1,\ldots,X_m, S_m, S_{m+1},\ldots)$ has the same probability law as that of $(X_{\pi(1)}, \ldots, X_{\pi(m)}, S_m, S_{m+1}, \ldots)$. Hence for each $k \leq m$

$$E[(X_1+\ldots+X_k)^p|F_{-m}] = E[(X_{\pi(1)}+\ldots+X_{\pi(k)})^p|F_{-m}]$$

and

$$E[Y_{-k}|F_{-m}] = E[\frac{1}{k}(X_1+\ldots+X_k)^p|F_{-m}]$$

$$= E[\frac{1}{m!}\sum_{\pi\in P_m}\frac{1}{k}(X_{\pi(1)}+\ldots+X_{\pi(k)})^p|F_{-m}]$$

$$\geq E[\frac{1}{m}(X_1+\ldots+X_m)^p|F_{-m}]$$

$$= Y_{-m}\ .$$

It is clear that $\{Y_{-m}\}$ is dominated by $E[X_1^p \mid F_{-m}]$. Thus

$\{Y_{-m}\}$ is uniformly integrable and converges almost surely in L_1. That is, $Y_{-m} \to X$ a.s. If $X > 0$ on some non-null set A, then $\lim_{m\to\infty} Y_{-2m} = X$ implies that $(S_{2m}/S_m)^p \to 2$. Hence, on A,

$$\frac{X_{m+1}+\ldots+X_{2m}}{X_1+\ldots+X_m} \to 2^{1/p} - 1 > 1$$

This is a contradiction, since the probability law of X does not change under permutation. That is, $(1, \ldots, m) \to (m+1, \ldots, 2m)$. ///

Remark: The convergence of Y_{-m} is also a consequence of a sub-additive theorem by Kingman, (1976).

The next strong convergence result is the almost sure convergence of the series $\Sigma_{n=1}^{\infty} a_n X_n$ where $\{X_n\}$ are exchangeable. The following result, Lemma 3.2.7, is a version of the Kolmogorov inequality for exchangeable random variables. The result is needed to prove the almost sure convergence of the series $\Sigma_{n=1}^{\infty} a_n X_n$.

Lemma 3.2.7 Let $X_1, \ldots, X_n$ be exchangeable random variables, with $EX_1^2 < \infty$. Let $S_k = \Sigma_{i=1}^{k} a_i X_i$, $k = 1, \ldots, n$, where $a_1, \ldots, a_n$ are real constants. Then for every $\lambda > 0$,

$$P[\max_{1\le k\le n} |S_k| \ge \lambda] \le \frac{4}{\lambda^2} (EX_1^2 - EX_1X_2) \Sigma_{i=1}^{n} a_i^2 + 2 \frac{E|X_1|}{\lambda} \Sigma_{i=1}^{n} |a_i|.$$

Proof: Let $F_k = \sigma\{X_1, \ldots, X_k, X_{k+1}+\ldots+X_n\}$, $k = 0, 1, \ldots, n-1$. Put $Y_k = a_k(X_k - E(X_k|F_{k-1}))$, and write $T_k=\Sigma_{i=1}^{k} Y_i$, $k=1,\ldots,n$. Then $\{Y_k\}$ are martingale differences, and $\{T_k, F_k\}_{k=1}^{n}$ is a square-integrable martingale. Thus, using the Hayek-Renyi inequality [Chow and Teicher (1978), p. 243 or Hall & Heyde (1980) Corollary 2.1, p. 14] with $p = 2$, for every $\lambda > 0$

$$P[\max_{1\le k\le n} |T_n| \ge \lambda] \le \lambda^{-2} \Sigma_{k=1}^{n} EY_k^2. \qquad (3.2.1)$$

Thus,

$$P[\max_{1\le k\le n} |S_k| \ge \lambda] = P[\max_{1\le k\le n} |\Sigma_{i=1}^{k} a_i X_i| \ge \lambda]$$

$$\le P[\max_{1\le k\le n} |\Sigma_{i=1}^{k} a_i (X_i - E(X_i|F_{i-1}))| \ge \frac{\lambda}{2}]$$

$$+ P[\max_{1\le k\le n} |\Sigma_{i=1}^{k} a_i E(X_i|F_{i-1})| \ge \frac{\lambda}{2}]$$

$$= P[\max_{1\le k\le n} |T_k| \ge \frac{\lambda}{2}] + P[\max_{1\le k\le n} |\Sigma_{i=1}^{k} a_i E(X_i|F_{i-1})| \ge \frac{\lambda}{2}]. \qquad (3.2.2)$$

By (3.2.1)

$$P[\max_{1\le k\le n} |T_k| \ge \frac{\lambda}{2}] \le \frac{4}{\lambda^2} \Sigma EY_k^2$$

$$= \frac{4}{\lambda^2} \Sigma_{k=1}^{n} a_k^2 E(X_k - E(X_k|F_{k-1}))^2$$

$$= \frac{4}{\lambda^2} \Sigma_{k=1}^{n} a_k^2 (EX_k^2 - 2EX_k E(X_k|F_{k-1}) + E(E^2(X_k|F_{k-1})))$$

$$= \frac{4}{\lambda^2} EX_1^2 \Sigma_{k=1}^{n} a_k^2 - \frac{8}{\lambda^2} \Sigma_{k=1}^{n} a_k^2 E[X_k E(X_k|F_{k-1})]$$

$$+ \frac{4}{\lambda^2} \Sigma_{k=1}^{n} a_k^2 EE[X_k E(X_k|F_{k-1})|F_{k-1}]$$

$$= \frac{4}{\lambda^2} EX_1^2 \Sigma_{k=1}^{n} a_k^2 - \frac{8}{\lambda^2} \Sigma_{k=1}^{n} a_k^2 E(X_k E(X_k|F_{k-1})) + \frac{4}{\lambda^2} \Sigma_{k=1}^{n} a_k^2 E(X_k E(X_k|F_{k-1}))$$

$$= \frac{4}{\lambda^2} EX_1^2 \Sigma_{k=1}^{n} a_k^2 - \frac{4}{\lambda^2} \Sigma_{k=1}^{n} a_k^2 E(X_k E(X_k|F_{k-1})). \qquad (3.2.3)$$

Noting that $E(X_k|F_{k-1}) = \frac{1}{n-k+1} \Sigma_{j=k}^{n} X_j$ and substituting in

(3.2.3), it follows that

$$\frac{4}{\lambda^2}[EX_1^2 \Sigma_{k=1}^n a_k^2 - \Sigma_{k=1}^n a_k^2 E(X_k E(X_k|F_{k-1}))]$$

$$= \frac{4}{\lambda^2}[EX_1^2 \Sigma_{k=1}^n a_k^2 - \Sigma_{k=1}^n a_k^2 E(X_k \frac{1}{n-k+1} \Sigma_{j=k}^n X_j)]$$

$$= \frac{4}{\lambda^2}[EX_1^2 \Sigma_{k=1}^n a_k^2 - \Sigma_{k=1}^n a_k^2 \left(\frac{EX_k^2 + (n-k) EX_1X_2}{n-k+1}\right)]$$

$$= \frac{4}{\lambda^2}[EX_1^2 \Sigma_{k=1}^n a_k^2 [1 - \frac{1}{n-k+1}] - EX_1X_2 \Sigma_{k=1}^n a_k^2 \frac{(n-k)}{n-k+1}]$$

$$= \frac{4}{\lambda^2}[EX_1^2 \Sigma_{k=1}^n a_k^2 \frac{(n-k)}{n-k+1} - EX_1X_2 \Sigma_{k=1}^n a_k^2 \frac{(n-k)}{n-k+1}]$$

$$= \frac{4}{\lambda^2}(EX_1^2 - EX_1X_2) \Sigma_{k=1}^n \frac{n-k}{n-k+1} a_k^2$$

$$\leq \frac{4}{\lambda^2} (EX_1^2 - EX_1X_2) \Sigma_{k=1}^n a_k^2, \text{ since } EX_1^2 - EX_1X_2 \geq 0.$$

Hence,

$$P[\max_{1\leq k\leq n} |T_k| \geq \frac{\lambda}{2}] \leq \frac{4}{\lambda^2}(EX_1^2 - EX_1X_2) \Sigma_{k=1}^n a_k^2. \qquad (3.2.4)$$

By Markov's inequality,

$$P[\max_{1\leq k\leq n} | \Sigma_{i=1}^k a_i E(X_i|F_{i-1})| \geq \frac{\lambda}{2}] \leq \frac{2}{\lambda} E|X_1| \Sigma_{k=1}^n |a_k|. \quad (3.2.5)$$

Substituting (3.2.4) and (3.2.5) into (3.2.2), the result follows.

///

Remark: Note that zero means are not assumed in Theorem 3.2.7 nor were they assumed in Theorem 3.2.6.

Theorem 3.2.8 Let $\{X_n\}$ be a sequence of exchangeable random variables with $EX_1^2 < \infty$. Let $\{a_n\}$ be a sequence of real constants such that $\Sigma_{n=1}^{\infty}|a_n| < \infty$. Then $\Sigma_{n=1}^{\infty} a_n X_n$ converges almost surely.

Proof: We follow the proof of Theorem 2, p. 110 of Tucker (1967). Note that for a sequence of numbers $\{x_n\}$,

$$\sup_{n \geq m} |x_n - x_m| = \max\{\sup(x_n - x_m), -\inf(x_n - x_m)\} \leq \frac{\varepsilon}{2}$$

$$\text{iff } \sup_{n \geq m}(x_n - x_m) \leq \frac{\varepsilon}{2} \text{ and } \inf_{n \geq m}(x_n - x_m) \geq \frac{-\varepsilon}{2}$$

$$\text{iff } \sup_{n \geq m} x_n \leq x_m + \frac{\varepsilon}{2} \text{ and } \inf_{n \geq m} x_n \geq x_m - \frac{\varepsilon}{2}, \text{ which implies}$$

$$\sup_{n \geq m} x_n - \inf_{n \geq m} x_n \leq \varepsilon, \text{ which implies } \overline{\lim}\, x_n - \underline{\lim}\, x_n \leq \varepsilon.$$

Putting $S_n = \Sigma_{i=1}^{n} a_i X_i$, we have

$$[\overline{\lim}\, S_n - \underline{\lim}\, S_n \geq 2\varepsilon] \subset [\sup_{n \geq m} |S_n - S_m| \geq \varepsilon].$$

But, $P[\sup_{n \geq m} |S_n - S_m| \geq \varepsilon] = \lim_{N \to \infty} P[\sup_{m < n \leq N} |S_n - S_m| \geq \varepsilon]$

$$\leq \lim_{N \to \infty} \{\frac{4}{\varepsilon^2}(EX_1^2 - EX_1X_2)\, \Sigma_{i=m+1}^{N} a_i^2 + \frac{2E|X_1|}{\varepsilon}\, \Sigma_{i=m+1}^{N} |a_i|\}$$

$$= \frac{4}{\varepsilon^2}(EX_1^2 - EX_1X_2)\, \Sigma_{i=m+1}^{\infty} a_i^2 + \frac{2E|X_1|}{\varepsilon}\, \Sigma_{i=m+1}^{\infty} |a_i| < \infty,$$

since clearly $\Sigma_1^{\infty} a_i^2 < \infty$. But letting $m \to \infty$ we get

$$P[\overline{\lim}\, S_n - \underline{\lim}\, S_n \geq 2\varepsilon] \leq \lim_{m \to \infty} P[\sup_{n \geq m} |S_n - S_m| \geq \varepsilon] = 0.$$

By Lemma 3.2.7 $\lim_{\lambda\to\infty} P[\sup_n |S_n| \geq \lambda]$

$$\leq \lim_{\lambda\to\infty} [\frac{4}{\lambda^2}(EX_1^2 - EX_1X_2)\,\Sigma_{i=1}^{\infty} a_i^2 + 2\,\frac{E|X_1|}{\lambda}\,\Sigma_{i=1}^{\infty} |a_i|]$$

$= 0$. Thus, $\overline{\lim}\, S_n$ and $\underline{\lim}\, S_n$ are a.s. finite.

Hence, $S_n = \Sigma_{i=1}^{n} a_i X_i$ converges a.s. ///

For the centered i.i.d. case $\Sigma_{k=1}^{\infty} a_k^2 < \infty$ is sufficient for a.s. convergence. That this is not so for exchangeable r.v.'s is shown by the exchangeable r.v.'s in Example 1.1.1 $X_n = Y_n + Y$ where $\{Y_n\}$ are i.i.d. and independent of Y. If $EY_n = 0 = EY$ and $EY_n^2 < \infty$ for all n and $EY^2 < \infty$, then $\Sigma_{n=1}^{\infty} \frac{1}{n} X_n = \Sigma_{n=1}^{\infty} \frac{1}{n} Y_n + Y\Sigma_{n=1}^{\infty} \frac{1}{n}$ does not converge a.s. unless $Y \equiv 0$ since the first term of the right hand side converges a.s. and $\Sigma_{n=1}^{\infty} \frac{1}{n} = \infty$

3.3 STRONG CONVERGENCE FOR TRIANGULAR ARRAYS OF EXCHANGEABLE RANDOM VARIABLES

The main result of this section uses an array of random variables such that $\{X_{nk} : 1 \leq k \leq n, n \geq 1\}$ is a finite exchangeable sequence for each n. This result is a strong law of large numbers that will incorporate a second moment condition and a condition on the measure of nonorthogonality for the random variables. In addition to this, the strong law of large numbers will be obtained using martingale methods and an appropriate choice of σ-fields. Finally, an example illustrating the main result of Section 3.3 will be presented. Let

$$F_{nn} = \sigma\{\Sigma_{k=1}^{n} X_{nk},\ \Sigma_{k=1}^{n+1} X_{(n+1),k},\ \ldots\} \qquad (3.3.1)$$

Recall from Property 1.2.4 that

$$\frac{1}{n}\Sigma_{k=1}^{n} X_{nk} = E(X_{n1}|F_{nn}) \text{ a.s.} \qquad (3.3.2)$$

Theorem 3.3.1 Let $\{X_{nk} : 1 \leq k \leq n, n \geq 1\}$ be a triangular array of random variables that are row-wise exchangeable. Let F_{nn} be the σ-field defined in (3.3.1). If

(i) $EX_{n1}X_{n2} \to 0$ a.s. $n \to \infty$,

(ii) $EX_{n1}^2 = o(n)$, and

(iii) $\{E(X_{n1}|F_{nn})\}_{n=1}^{\infty}$ is a reverse martingale, then

$\frac{1}{n} \Sigma_{k=1}^{n} X_{nk} \to 0$ a.s.

Proof: By (3.3.2) and the reverse martingale inequality (rf: Lemma 4.3.1)

$$P[\sup_{n\geq m} |\frac{1}{n} \Sigma_{k=1}^{n} X_{nk}| > \varepsilon] = P[\sup_{n\geq m} |E(X_{n1}|F_{nn})| > \varepsilon]$$

$$\leq \frac{1}{\varepsilon} \int_{[\sup_{n\geq m}|E(X_{n1}|F_{nn})| > \varepsilon]} |E(X_{m1}|F_{mm})| dP$$

$$\leq \frac{1}{\varepsilon} \int_{\Omega} |E(X_{m1}|F_{mm})| dP$$

$$= \frac{1}{\varepsilon} \int_{\Omega} |\frac{1}{m} \Sigma_{k=1}^{m} X_{mk}| dP.$$

By Holder's inequality with $p = q = 2$,

$$\int_{\Omega} |\frac{1}{m} \Sigma_{k=1}^{m} X_{mk}| dP \leq [\int_{\Omega} (\frac{1}{m} \Sigma_{k=1}^{m} X_{mk})^2 dP]^{1/2}$$

$$= [\int_{\Omega} (\frac{1}{m^2} \Sigma_{k=1}^{m} X_{mk}^2 + \frac{1}{m^2} \Sigma\Sigma_{k\neq\ell} X_{m\ell}X_{mk}) dP]^{1/2}$$

$$= [E(\frac{1}{m^2} \Sigma_{k=1}^{m} X_{mk}^2 + \frac{1}{m^2} \Sigma\Sigma_{k\neq\ell} X_{mk}X_{m\ell})]^{1/2}.$$

By exchangeability,

$$[E(\frac{1}{m^2}\Sigma_{k=1}^m X_{mk}^2 + \frac{1}{m^2}\underset{k\neq\ell}{\Sigma\Sigma} X_{mk}X_{m\ell})]^{1/2}$$

$$= [\frac{EX_{m1}^2}{m} + \frac{m(m-1)}{m^2} EX_{m1}X_{m2}]^{1/2}.$$

Thus,

$$\frac{1}{\varepsilon}\int_\Omega |\frac{1}{m}\Sigma_{k=1}^m X_{mk}| dP \leq \frac{1}{\varepsilon}[\frac{1}{m}EX_{m1}^2 + \frac{m(m-1)}{m^2}EX_{m1}X_{m2}]^{1/2}$$

which converges to 0 as $m \to \infty$. ///

The following example provides situations where the reverse martingale assumption on $\{E(X_{n1}|F_{nn})\}^\infty$ is satisfied. A triangular array is formed by subtracting the sample mean from independent, identically distributed random variables.

Example 3.3.1 Let $Y_1, \ldots, Y_n \ldots$ be independent, identically distributed random variables with mean μ and variance σ^2. Define

(i) $X_{nj} = Y_j - \bar{Y}_n$ where $\bar{Y}_n = \frac{1}{n}\Sigma_{i=1}^n Y_i$,

(ii) $F_n = \sigma\{\Sigma_{i=1}^n Y_i, \Sigma_{i=1}^{n+1} Y_i, \ldots\}$.

It is interesting to note that the σ-field $\sigma\{\Sigma_{i=1}^n X_{ni}, \Sigma_{i=1}^{n+1} X_{(n+1),i}, \ldots\} = \sigma\{0,0,\ldots\}$ in this setting. Recall from (3.3.2) that

(i) $\bar{Y}_n = E(Y_1|F_n)$ a.s. (3.3.3)

(ii) $\bar{Y}_{n+1} = E(Y_1|F_{n+1})$ a.s. (3.3.4)

Problem 1.4 established that $\{X_{nk} : 1 \leq k \leq n\}$ is exchangeable for each n. The next three propositions will show that (1) $\rho_n = EX_{n1}X_{n2} \to 0$ as $n \to \infty$ (2) $\{E(X_{n1}|F_n)\}_{n=1}^{\infty}$ is a reverse martingale and (3) $\{E(n/(n-1)X_{n1}^2|F_n)\}_{n=1}^{\infty}$ is a reverse martingale.

Proposition 3.3.2 Let $\{X_{nk} : 1 \leq k \leq n, n \geq 1\}$ be the array of random variables in Example 3.3.1. Then $EX_{n1}X_{n2} \to 0$ as $n \to \infty$.

Proof: Since $\{Y_n\}$ are independent, and identically distributed

$$EX_{n1}X_{n2} = E[(Y_1 - \bar{Y}_n)(Y_2 - \bar{Y}_n)]$$

$$= E(Y_1Y_2) - E(Y_1\bar{Y}_n) - E(Y_2\bar{Y}_n) + E(\bar{Y}_n)^2$$

$$= [EY_1]^2 - \frac{(n-1)}{n}\mu^2 - 2\frac{EY_1^2}{n} - \frac{(n-1)}{n}\mu^2 + \frac{EY_1^2}{n} + \frac{(n-1)}{n}\mu^2$$

$$= \mu^2 - \frac{(n-1)}{n}\mu^2 - \frac{EY_1^2}{n} \to 0 \text{ as } n \to \infty. \qquad ///$$

Proposition 3.3.3 Let $\{X_{nk} : 1 \leq k \leq n, n \geq 1\}$ be the array defined in Example 3.3.1. Then, $\{E(X_{n1}|F_n)\}_{n=1}^{\infty}$ is a reverse martingale.

Proof: Let $Z_n = E(X_{n1}|F_n)$, then $\sigma\{Z_{n+1}\} \subseteq F_{n+1} \subseteq F_n$ and

$$E(Z_n|Z_{n+1}) = E(E(X_{n1}|F_n)|Z_{n+1})$$

$$= E(X_{n1}|Z_{n+1}) \text{ a.s. since } F_n \supset \sigma\{Z_{n+1}\}.$$

Here $F_n \supseteq F_{n+1} \supseteq \sigma\{Z_{n+1}\}$.

Therefore, by (3.3.3) and (3.3.4)

$$E(X_{n1}|Z_{n+1}) = E(Y_1|Z_{n+1}) - E(\tfrac{1}{n}\Sigma_{i=1}^{n} Y_i|Z_{n+1})$$

$$= E(Y_1|Z_{n+1}) - E(E(Y_1|F_n)|Z_{n+1})$$

$$= E(Y_1|Z_{n+1}) - E(Y_1|Z_{n+1}) \qquad (3.3.5)$$

$$= E(Y_1|Z_{n+1}) - E(E(Y_1|F_{n+1})|Z_{n+1})$$

$$= E(Y_1|Z_{n+1}) - E(\tfrac{1}{n+1}\Sigma_{i=1}^{n+1} Y_i|Z_{n+1})$$

$$= E(Y_1 - \bar{Y}_{n+1}|Z_{n+1}).$$

Noting again that $\sigma\{Z_{n+1}\} \subseteq F_{n+1}$, it follows that

$$E(Y_1 - \bar{Y}_{n+1}|Z_{n+1}) = E(X_{(n+1),1}|Z_{n+1})$$

$$= E(E(X_{(n+1),1}|F_{n+1})|Z_{n+1})$$

$$= E(Z_{n+1}|Z_{n+1})$$

$$= Z_{n+1} \text{ a.s.} \qquad ///$$

<u>Remark</u>: $\{E(X_{n1}|F_n)\}$ is a reverse martingale difference sequence, since from (3.3.5) in the above proof, $E(Z_n|Z_{n+1}) = Z_{n+1} = 0$ a.s. Thus $E(X_{n1}|F_n) = 0$ a.s. Since $\{E(X_{n1}|F_n)\}$ is a reverse martingale, condition (iii)

of Theorem 3.3.1 is satisfied. It is more interesting to show that $E(\frac{n}{n-1} X_{n1}^2 | F_n)$ is a reverse martingale.

Proposition 3.3.4 Let $\{X_{nk} : 1 \leq k \leq n, n \geq 1\}$ be the array defined in Example 3.3.1. Then $\{E(\frac{n}{n-1} X_{n1}^2 | F_n)\}_{n=1}^{\infty}$ is a reverse martingale.

Proof: It suffices to show that

$$E(\frac{n}{n-1} E(X_{n1}^2 | F_n) | F_{n+1}) = \frac{n+1}{n} E(X_{n+1}^2 | F_{n+1}) \text{ a.s.}$$

First,

$$\begin{aligned} E(X_{n1}^2 | F_n) &= E[(Y_1 - \bar{Y}_n)^2 | F_n] \\ &= E(Y_1^2 - 2Y_1\bar{Y}_n + \bar{Y}_n^2 | F_n) \\ &= E(Y_1^2 | F_n) - 2E(Y_1\bar{Y}_n | F_n) + E(\bar{Y}_n^2 | F_n). \end{aligned}$$

Since $\sigma\{\bar{Y}\} \in F_n$ and by (3.3.2)

$$\begin{aligned} -2E(Y_1\bar{Y}_n | F_n) + E(\bar{Y}_n^2 | F_n) &= -2\bar{Y}_n E(Y_1 | F_n) + \bar{Y}_n E(\bar{Y}_n | F_n) \\ &= -2\bar{Y}_n\bar{Y}_n + \bar{Y}_n\bar{Y}_n \\ &= -2\bar{Y}_n^2 + \bar{Y}_n^2 \\ &= -\bar{Y}_n^2 \quad \text{a.s.} \end{aligned}$$

Hence,

$$E(Y_1^2|F_n) - 2E(Y_1Y_n|F_n) + E(\bar{Y}_n^2|F_n)$$

$$= E(Y_1^2|F_n) - \bar{Y}_n^2$$

$$= E(Y_1^2|F_n) - E(\bar{Y}_n^2|F_n)$$

$$= E(Y_1^2|F_n) - \frac{1}{n^2}(\sum_i \sum_j E(Y_iY_j|F_n)).$$

Thus,

$$E[E[(Y_1 - \bar{Y}_n)^2|F_n]|F_{n+1}] = E(E(Y_1^2|F_n)|F_{n+1})$$

$$- \frac{1}{n^2} (\sum_i \sum_j E(E(Y_iY_j|F_n)|F_{n+1})$$

$$= E(Y_1^2|F_{n+1}) - \frac{1}{n^2} [nE(E(Y_1^2|F_n)|F_{n+1})]$$

$$- \frac{1}{n^2} [n(n-1)E(Y_1Y_2|F_n)|F_{n+1})]$$

$$= E(Y_1^2|F_{n+1}) - \frac{1}{n}E(Y_1^2|F_{n+1}) - \frac{(n-1)}{n} E(Y_1Y_2|F_{n+1})$$

$$= \frac{(n-1)}{n} E(Y_1^2|F_{n+1}) - \frac{(n-1)}{n} E(Y_1Y_2|F_{n+1}).$$

Similarly

$$E[(Y_1 - \bar{Y}_{n+1})^2|F_{n+1}] = \frac{n}{n+1} E(Y_1^2|F_{n+1}) - \frac{n}{n+1} E(Y_1Y_2|F_{n+1}).$$

Thus,

$$E[E[(\frac{n}{n-1})X_{n1}^2|F_n]|F_{n+1}] = E[E[(\frac{n}{n-1})(Y_1 - \bar{Y}_n)^2|F_n]|F_{n+1}]$$

$$= \frac{n}{n-1}((\frac{n-1}{n}) E(Y_1^2|F_{n+1}) - \frac{(n-1)}{n}E(Y_1Y_2|F_{n+1}))$$

$$= E(Y_1^2|F_{n+1}) - E(Y_1Y_2|F_{n+1})$$

$$= (\frac{n+1}{n}) (\frac{n}{n+1}) E(Y_1^2|F_{n+1}) - \frac{n}{n+1} E(Y_1Y_2|F_{n+1}))$$

$$= (\frac{n+1}{n}) E[(Y_1 - \bar{Y}_{n+1})^2|F_{n+1}]$$

$$= E(\frac{(n+1)}{n} X^2_{(n+1),1}|F_{n+1}) \quad \text{a.s.}$$

Hence, $\{E(\frac{n}{n-1} X^2_{n1}|F_n)\}_{n=1}^{\infty}$ is a reverse martingale. ///

By Proposition 3.3.4 and the reverse martingale inequality,

$$P[\sup_{n\geq m} |\frac{1}{n} \Sigma^n_{k=1} X^2_{nk}| > \varepsilon] = P[\sup_{n\geq m} |E(X^2_{n1}|F_n)| > \varepsilon]$$

$$\leq P[\sup_{n\geq m} |E(\frac{n}{n-1} X^2_{n1}|F_n)| > \varepsilon]$$

$$\leq \frac{1}{\varepsilon} \int_\Omega |E(\frac{m}{m-1} X^2_{m1}|F_m)| \, dP. \qquad (3.3.6)$$

Using (3.3.6) and Proposition 3.3.4, it can be shown that the weighted sum of the squared deviations from the mean of a random sample converges a.s. to zero. The proof is similar to that of Theorem 3.3.1 and will not be listed.

Proposition 3.3.5 Let X_{nk} be the array in Example 3.3.1, where $X_{nk} = Y_k - \bar{Y}_n$. Let $J_{nn} = \sigma\{\Sigma^n_{k=1} X_{nk}, \Sigma^n_{k=1} X^2_{nk}, \Sigma^{n+1}_{k=1} X_{(n+1),k}, \Sigma^n_{k=1} X^2_{(n+1)k}, \ldots\}$. Then for $\alpha > 1$

$$\Sigma^n_{k=1} (Y_k - \bar{Y}_n)^2/n^\alpha \to 0 \quad \text{a.s.}$$

The next result is a strong law of large numbers for infinite arrays of real-valued random variables. In fact, complete convergence will be obtained rather than almost sure convergence. A more detailed proof will be given for the Banach space version of the next result (rf: Theorem 6.2.3). Hence, the proof will be omitted for the real-valued version.

Theorem 3.3.6 Let $\{X_{nk}\}$ be an array of real-valued random variables and let $\{X_{nk} : k \geq 1\}$ be exchangeable for each n. If for each $\eta > 0$

$$\Sigma_{n=1}^{\infty}\mu_n(\{P_\nu : |E_\nu(X_{n1})| > \eta\}) < \infty \tag{3.3.7}$$

and for some $q \geq 1$

$$\Sigma_{n=1}^{\infty}E(|X_{n1}|^{2q})/n^q < \infty, \tag{3.3.8}$$

then

$$|\frac{1}{n}\Sigma_{k=1}^{n}X_{nk}| \to 0 \text{ completely.}$$

Remark: As in (1.3.8), Condition (3.3.7) is related to the measure of nonorthogonality. That is, $\Sigma\,\mu_n(\{P_\nu : |E_\nu(X_{n1})| > \eta\}) < \infty$ implies that $\mu_n(\{P_\nu : |E_\nu(X_{n1})| > \eta\}) \to 0$ which is necessary, by (1.3.8), for the measure of nonorthogonality, ρ_n, to converge to zero.

3.4 PROBLEMS

3.1 Let $X_1, \ldots, X_n$ be random variables with $E|X_k| < \infty$ for all k. Let $F_k = \sigma\{X_1, \ldots, X_k, X_{k+1}+\ldots+X_n\}$. Show that $E[E(X_k|F_{k-1})]^2 = E[X_kE(X_k|F_{k-1})]$ a.s.

3.2 Let $X_1, \ldots, X_t$ be random variables. Show that for each $\varepsilon > 0$

$$P[\Sigma_{i=1}^{t} |X_i| > \varepsilon] \leq \Sigma_{i=1}^{t} P[|X_i| > \frac{\varepsilon}{t}].$$

3.3 Let $X_1, \ldots, X_n$ be exchangeable random variables. Show that $EX_1^2 - EX_1X_2 \geq 0$.

3.4 Show that $E(|X|^2) < \infty$ implies that $E|X|^p < \infty$ for all $p \in [1,2)$.

3.5 Show that $P[|X_n| \geq t] \leq P[|X| \geq t]$ for each $t \in R$ implies that $E|X_n|^r \leq E|X|^r$ for each $r > 0$ such that $E|X|^r$ exists.

3.6 Show that $\Sigma_{n=1}^{\infty} E(X_{n1}X_{n2}) < \infty$ implies $\Sigma_{n=1}^{\infty} \mu_n(\{P_\nu : |E_\nu(X_{n1})| > \eta\}) < \infty$ but that the converse need not hold.

3.7 In the proof of Theorem 3.2.6, show that for $y_1, \ldots, y_m \geq 0$, $1 \leq k \leq m$, $0 < p < 1$, $\frac{1}{m}(y_1 + \ldots + y_m)^p$

$$\leq \frac{1}{m!} \sum_{\pi \varepsilon P_m} \frac{1}{k}(Y_{\pi(1)} + \ldots + Y_{\pi(k)})^p.$$

CHAPTER IV

PRELIMINARIES AND BACKGROUND MATERIAL FOR RANDOM ELEMENTS

4.0 INTRODUCTION

This chapter will briefly introduce the concepts and background material for the limit theorems of exchangeable random variables in Banach spaces which will be presented in Chapters 5 and 6. Many of the properties and results for random elements are the same as those for (real-valued) random variables. However, there are notable exceptions, several of which lead to very rich and extensive research areas. Unfortunately, only some of these will be mentioned with none being described in detail. Very detailed developments for random elements are in Taylor (1978), for geometric considerations of Banach spaces relating to probability results are in Woyczynski (1978), and for conditional expectations and martingale theory in Banach spaces are in Scalora (1961) and Diestel and Uhl (1977).

In Section 4.1 the definitions of random elements and exchangeability are provided for Banach spaces. Several measurability and expectation properties are developed, and the geometric consideration of Banach spaces which are B-convex, type p, or p-uniformly smooth are briefly described. Finally, the distribution concepts of tightness and tail bounded probabilities are given.

The weak convergence of probability measures is introduced in Section 4.2. A Gaussian random element which serves as a limit random element in the central limit theorem is also given. A version of de Finetti's theorem for random elements is provided, and measurability and distributional properties relating to exchangeable random elements are given.

In Section 4.3 conditional expectations of random elements are defined, and their properties are developed in a martingale setting. A crucial reverse martingale inequality is obtained, and the sequence of averages of exchangeable random elements are shown to be a reverse martingale. The chapter is concluded with several problems which further illustrate these concepts.

4.1 BASIC DEFINITIONS AND PROPERTIES OF EXCHANGEABLE RANDOM ELEMENTS

Let $(\Omega, \mathcal{A}, P)$ denote a probability space, let E denote a separable (real) Banach space with norm $\|\ \|$, let E* denote the dual space of all continuous, linear functionals on E, and let $\mathcal{B}(E)$ denote the σ-field of Borel subsets generated by the open subsets of E. A random element on E is a function, X, from Ω into E such that $X^{-1}(\mathcal{B}(E)) \subset \mathcal{A}$. Moreover, it can be shown (Taylor (1978), p. 27) that $X: \Omega \to E$ is a random element if and only if f(X) is a random variable for each $f \in E^*$.

Definition 4.1.1 A finite sequence of random elements $\{X_1, \ldots, X_n\}$ is said to be exchangeable if

$$P[X_1 \in B_1, \ldots, X_n \in B_n] = P[X_{\pi 1} \in B_1, \ldots, X_{\pi n} \in B_n]$$

for each permutation π of the indices {1, ..., n} and for all $B_1, \ldots, B_n \in \mathcal{B}(E)$. A collection of random elements is said to be exchangeable if every finite subset is exchangeable.

Separability of E is assumed to avoid measurability problems, but otherwise, properties for exchangeable random elements are similar to those of random variables. For example, all joint distributions of exchangeable random elements are the same, and a Borel function of E_1 into E_2 (both Banach spaces) preserves exchangeability (similar proofs to Properties 1.1.1 and 1.1.2). The following proposition will be used in applications in Chapter 6 and illustrates some of the exchangeability techniques.

<u>Proposition 4.1.1</u> Let $X_1, \ldots, X_n$ be exchangeable random elements. If $H_n(X_1, \ldots, X_n) = H_n$ is random variable and a Borel, symmetric function of $(X_1, \ldots, X_n)$, then $\{X_1/H_n, \ldots, X_n/H_n\}$ is exchangeable when $H_n \neq 0$.

<u>Proof</u>: The distribution of $(X_1, \ldots, X_n)$ is completely determined by the values of $E[f(X_1, \ldots, X_n)]$ where f is a bounded, Borel function from E^n into R(see Problem 4.10). For a bounded, Borel function f from E^n into R, let

$$g(x_1, \ldots, x_n) = f\left(\frac{x_1}{H(x_1,\ldots,x_n)}, \ldots, \frac{x_n}{H(x_1, \ldots, x_n)}\right).$$

Since $\{X_1, \ldots, X_n\}$ are exchangeable and H is symmetric,

$$E\left[f\left(\frac{X_1}{H(X_1,\ldots,X_n)}, \ldots, \frac{X_n}{H(X_1, \ldots, X_n)}\right)\right]$$

$$= E[g(X_1, \ldots, X_n)]$$

$$= E[g(X_{\pi 1}, \ldots, X_{\pi n})]$$

$$= E\left[f\left(\frac{X_{\pi 1}}{H_n}, \ldots, \frac{X_{\pi n}}{H_n}\right)\right]$$

for each permutation π of $\{1, 2, \ldots, n\}$. ///

Remark 4.1.1 The distribution of a random variable is completely determined by the values of its characteristic function

$$E[\exp(itX)] = E[\cos t X] + i E[\sin t X].$$

Since cos and sin are bounded, continuous functions, it is easy to see how Problem 4.10 would follow for random variables.

The concepts of independence and identical distributions for random elements parallel those of random variables (see Taylor (1978) for details). The major differences occur with integral conditions. The expected value EX of a random element X will be defined by a Bochner integral. Moreover, since E is separable, the Pettis integral and the Bochner integral will exist and agree when $E||X|| < \infty$, and $f(EX) = E[f(X)]$ for each $f \in E^*$.

Mourier (1953) proved that for independent, identically distributed random elements $\{X_n\}$

$$\frac{1}{n} \Sigma_{k=1}^{n} X_k \to EX_1 \quad \text{a.s.}$$

if and only if $E||X_1|| < \infty$. Beck (1963) obtained a SLLN for independent random elements with uniformly bounded, second centered absolute moments by restricting the Banach space to be B-convex.

Definition 4.1.1 A normed linear space X is said to be convex of type (B) if there is an integer $t > 0$ and an $\varepsilon > 0$ such that for all $x_1, \ldots, x_t \in X$ with $||x_i|| \leq 1$, $i = 1, \ldots, t$, then

$$||\pm x_1 \pm x_2 \pm \ldots \pm x_t|| < t\,(1 - \varepsilon) \qquad (4.1.1)$$

for some choice of + and - signs.

Geometric considerations relating to laws of large numbers for Banach spaces were further developed by Hoffmann-Jørgensen and Pisier (1976). Let $\{Y_n\}$ be a Bernoulli sequence, that is, $\{Y_n\}$ are independent random variables with $P[Y_n = +1] = \frac{1}{2} = P[Y_n = -1]$ for each n. Let $E^\infty = E \times E \times E \ldots$ and define

$$C(E) = \{(x_n) \in E^\infty : \Sigma_{n=1}^\infty Y_n x_n \text{ converges in probability}\}.$$

Then a separable Banach space X is said to be of <u>type p</u>, $1 \leq p \leq 2$, if and only if there exists $A \in R^+$ such that

$$E||\Sigma_{n=1}^\infty Y_n x_n||^p \leq A\, \Sigma_{n=1}^\infty ||x_n||^p$$

for all $(x_n) \in C(E)$. The following results and accompanying remarks are from Hoffmann-Jørgensen and Pisier (1976).

<u>Theorem 4.1.2</u> Let $1 \leq p \leq 2$, then the following four statements are equivalent:

(i) The separable Banach space E is of type p.

(ii) There exists $A \in R^+$ such that

$$E||\Sigma_{k=1}^n X_k||^p \leq A\, \Sigma_{k=1}^n E||X_k||^p \qquad (4.1.2)$$

for all independent random elements $X_1, \ldots, X_n$ with mean 0 and finite pth moments.

(iii) The strong law of large numbers,

$$\frac{1}{n} \Sigma_{k=1}^n X_k \to 0 \text{ with probability one,}$$

holds for all sequences of independent random elements $\{X_n\}$ such that $EX_n = 0$ for all n and such that Chung's condition holds,

$$\Sigma_{n=1}^{\infty} \frac{E||X_n||^p}{n^p} < \infty.$$

(iv) If

$$\Sigma_{n=1}^{\infty} \frac{||x_n||^p}{n^p} < \infty$$

and $\{Y_n\}$ is a Bernoulli sequence, then

$$\frac{1}{n} \Sigma_{k=1}^{n} Y_k x_k \to 0$$

in probability.

Type p is often called Rademacher type p. From the triangle inequality for the norm, it is easy to see that every separable Banach space is of type 1. Clearly, R with the usual Euclidean norm is of type 2. Every separable Hilbert space and every finite-dimensional Banach space is of type 2. If E is of type p, then E is of type q for all $q \leq p$. Hence, for $p \geq 1$, ℓ^p and L^p are of type min{2,p}. Pisier (1974) showed that E is B-convex if and only if E is of type p for some p > 1.

A space is said to be <u>p-uniformly smooth</u> if the modulus of smoothness

$$\rho(t) = \sup\{\tfrac{1}{2}(||x+y||+||x-y|| - 2): ||x||=1 \text{ and } ||y|| = t\}$$

satisfies $\rho(t) = o(t^p)$ as $t \to 0$. A space is said to be p-uniformly smoothable if there exists an equivalent norm which makes the space p-uniformly smooth. Thus, p-uniformly smooth spaces are superreflexive, but James (1974) has given an example of a B-convex space (and hence type p for some p > 1) which is not reflexive and thus not p-uniformly smooth.

The necessity of these geometric considerations is easy to illustrate. An alternative approach is to require tightness of the probability measures. The random elements $\{X_\alpha: \alpha \in \Gamma\}$ are said to be tight if there exists a compact set $K_\varepsilon \subset E$ such that $P[X_\alpha \in K_\varepsilon] > 1 - \varepsilon$ for all $\alpha \in \Gamma$. Since each probability measure on a complete separable metric space is tight (Billingsley (1968) p. 10), identically distributed random elements in a separable Banach space are tight. Using tightness, Taylor and Wei (1978) and (1979) were able to obtain laws of large numbers for independent and weakly uncorrelated random elements without assuming geometric conditions. Random elements $\{X_\alpha: \alpha \in \Gamma\}$ are said to be weakly uncorrelated if $\{f(X_\alpha): \alpha \in \Gamma\}$ are uncorrelated random variables for each $f \in E^*$. A very general condition involving tightness and moment conditions is provided by Definition 4.1.2.

Definition 4.1.2 The random elements $\{X_\alpha: \alpha \in \Gamma\}$ are said to satisfy condition (T) if for each $\varepsilon > 0$ there exists a compact subset $K_\varepsilon \subset E$ such that

$$E||X_\alpha I_{[X_\alpha \notin K_\varepsilon]}|| < \varepsilon \quad \text{for all } \alpha \in \Gamma. \qquad (4.1.3)$$

Problem 4.4 relates condition (T) to uniform integrability, and condition (T) is implied by tightness and uniformly bounded pth moments, $p > 1$. Finally, random elements $\{X_\alpha: \alpha \in \Gamma\}$ are said to have uniformly bounded tail probabilities (denoted $\{X_\alpha\} << X$) by a random variable X if $P[||X_\alpha|| \geq t] \leq P[|X| \geq t]$ for all $t \in R$ and $\alpha \in \Gamma$. Clearly, identically distributed random elements have uniformly bounded tail probabilities, and uniformly bounded pth moments, $p > 0$ imply uniformly bounded tail probabilities.

4.2 CONVERGENCE OF MEASURES AND DE FINETTI'S THEOREM

In Section 1.3 the topology of convergence of distribution functions was discussed. The discussion of weak convergence of probability measures will parallel the weak convergence of distribution functions. A more extensive, detailed development is in Billingsley (1968). In a Banach space the concept of distribution functions is not available, but the distribution of a probability measure can be described by its values on the Borel subsets. A sequence of probabilities $\{P_n\}$ on $\mathcal{B}(E)$ is said to <u>converge weakly</u> to P (denoted by $P_n \to P$) if $\int f dP_n \to \int f dP$ for every bounded continuous function $f: E \to R$. It is easy to see that when E = R weak convergence coincides with convergence in distribution. Portmanteau Theorem (Billingsley (1968) pp. 11-14) provides 5 equivalent conditions relating to weak convergence.

<u>Theorem 4.2.1</u> Let S be a metric space and let P_n, P be probability measures on $(S, \mathcal{B}(S))$. The following five conditions are equivalent:

(i) $P_n \to P$.

(ii) $\lim_n \int f dP_n = \int f dP$ for all bounded, uniformly continuous real f.

(iii) $\limsup_n P_n(F) \leq P(F)$ for all closed F.

(iv) $\liminf_n P_n(G) \geq P(G)$ for all open G.

(v) $\lim_n P_n(A) = P(A)$ for all P-continuity sets A.

<u>Remark</u>: The set $A \in \mathcal{B}(S)$ is said to be a P-continuity set if P(boundary of A) = 0.

A sequence of random elements $\{X_n\}$ on E is said to

<u>converge in distribution</u> (denoted $X_n \xrightarrow{\mathcal{D}} X$) to X if and only if $P_{X_n} \rightarrow P_X$. The following 5 equivalent conditions follows immediately from Theorem 4.2.1.

(i) $X_n \xrightarrow{\mathcal{D}} X$.

(ii) $\lim_n E[f(X_n)] = E[f(X)]$ for all bounded , uniformly continuous real f.

(iii) $\limsup_n P[X_n \in F] \leq P[X \in F]$ for all closed F.

(iv) $\liminf_n P[X_n \in G] \geq P[X \in G]$ for all open G.

(v) $\lim_n P[X_n \in A] = P[X \in A]$ for all X-continuity sets A.

When considering $E = C[0,1]$, a necessary and sufficient condition for random elements $\{X_n\}$ to converge in distribution to X is that the distribution functions $F_{X_n(t_1),\ldots,X_n(t_k)}$ converge to $F_{X(t_1),\ldots,X(t_k)}$ for each choice of $t_1, \ldots, t_k \in [0,1]$ and for the sequence $\{X_n\}$ to be tight (rf: Billingsley (1968) p.40).

In the development of central limit theorems in separable Banach spaces, Gaussian random elements will serve as the limit random elements and characteristic functionals will be substituted for the characteristic functions.

<u>Definition 4.2.1</u> The characteristic functional ϕ: $E^* \rightarrow$ complex numbers of a probability measure μ on E is given by $\phi(f) = \int_E \exp(if(x))\mu(dx)$, $f \in E^*$.

The characteristic functional uniquely determines the probability measure μ (see Vakhania (1981) Section (4.1) for properties). If $\int_E f^2(x)\mu(dx) < \infty$ for all $f \in E^*$, the covariance operator H: $E^* \rightarrow E$ is the bounded, linear, symmetric and non-negative operator determined by the bilinear form $\int_E f(x)g(x)\mu(dx)$ by posing $\langle Hf,g\rangle$

$= \int_E f(x)g(x)\mu(dx)$, $f,g \in E^*$. If μ is the distribution with characteristic functional of the form $\exp(-\frac{1}{2}\langle Hf,f\rangle)$, $f \in E^*$, then μ is said to be a <u>Gaussian distribution</u>.

Olshen (1974) showed that if $\{X_n : n \geq 1\}$ are exchangeable random variables taking values in a complete, separable metric space then there exists a real-valued random variable M such that $\{X_n\}$ are conditionally independent and identically distributed given M. Thus, following the development in Section 1.3, let F be the collection of probability measures on $\mathcal{B}(E)$ and let F be equipped with the σ-field generated by the topology of weak convergence of the probability measures. If $\{X_n\}$ is an infinite sequence of exchangeable random elements in E, then there exists a probability measure on $\mathcal{B}(F)$ such that (imprecisely stated)

$$P(B) = \int_F P_\nu(B)d\mu(P_\nu) \tag{4.2.1}$$

for any $B \in \sigma\{X_n : n \geq 1\}$ and $P_\nu(B)$ is calculated as if the random elements $\{X_n\}$ are i.i.d. with respect to P_ν.

Again, it is important to recall from Examples 1.2.1 and 1.3.1 that conditions on the conditional means $E_\nu(X_n)$ are the key requirements in obtaining limit theorems for exchangeable random elements. Moreover, the sets in (1.3.6) and (1.3.7) retain their measurability properties when the absolute value is replaced by the norm of the Banach space. Infinite exchangeability is again essential in establishing (4.2.1), but Theorem 4.2.2 provides a framework for considering finite exchangeable random elements.

<u>Theorem 4.2.2</u> Let $\{X_{nk} : 1 \leq k \leq n, n \geq 1\}$ be an array of random elements which are exchangeable for each n. If for each k, $X_{nk} \xrightarrow{r} X_{\infty k}$, $(E||X_{nk} - E_{\infty k}||^r \to 0)$ for some $r > 0$, then the infinite sequence $\{X_{\infty k} : k \geq 1\}$ is exchangeable.

Proof: Consider the set $\{X_{n1}, \ldots, X_{nk}\} \subset \{X_{n1}, \ldots, X_{mn}\}$ and the vector $(X_{n1}, \ldots, X_{nk})$. If $X_{ni} \xrightarrow{r} X_{\infty i}$, $1 \leq i \leq k$, then by Problem 4.6 $(X_{n1}, \ldots, X_{nk}) \xrightarrow{r} (X_{\infty 1}, \ldots, X_{\infty k})$. Hence, $(X_{n1}, \ldots, X_{nk}) \xrightarrow{\mathcal{D}} (X_{\infty 1}, \ldots, X_{\infty k})$. For all $P_{X_{\infty 1}}$-continuity sets $A_i \in \mathcal{B}(E)$, $1 \leq i \leq k$,

$$P[X_{n1} \in A_1, \ldots, X_{nk} \in A_k] \to P[X_{\infty 1} \in A_1, \ldots, X_{\infty k} \in A_k];$$

by Theorem 4.2.1. By exchangeability,

$$P[X_{n1} \in A_1, \ldots, X_{nk} \in A_k] = P[X_{n\pi 1} \in A_1, \ldots, X_{n\pi k} \in A_k]$$

for each permutation π of $(1, \ldots, k)$. Hence,

$$P[X_{n\pi 1} \in A_1, \ldots, X_{n\pi k} \in A_k] \to P[X_{\infty\pi 1} \in A_1, \ldots, X_{\infty\pi k} \in A_k]$$

as $n \to \infty$ for all $A_1, \ldots, A_k$ which are $P_{X_{\infty 1}}$-continuity sets. Since the limits are unique, and the $P_{X_{\infty 1}}$-continuity sets form a determining class, it follows that

$$P[X_{\infty 1} \in B_1, \ldots, X_{\infty k} \in B_k] = P[X_{\infty\pi 1} \in B_1, \ldots, X_{\infty\pi k} \in B_k]$$

for all $(B_1, \ldots, B_k) \in \mathcal{B}(X^k)$. Thus, $\{X_{\infty 1}, \ldots, X_{\infty k}\}$ and $\{X_{\infty\pi 1}, \ldots, X_{\infty\pi k}\}$ have identical joint distributions. Hence, the sequence $\{X_{\infty k}: k \geq 1\}$ is exchangeable. ///

The next result shows that if $\{X_{nk}: 1 \leq k \leq n, n \geq 1\}$ are row-wise exchangeable random elements which converge in the second mean for each k and $E[f(X_{n1})f(X_{n2})] \to 0$ as $n \to \infty$, then $E[f(X_{\infty 1})f(X_{\infty 2})] = 0$ and $E_\nu(X_{\infty 1}) = 0$ μ-almost surely where E_ν is the expectation with respect to P_ν. Here $X_{nk} \overset{2}{\to} X_{\infty k}$ as $n \to \infty$, and the resulting sequence $\{X_{\infty k}: k \geq 1\}$ consists of independent, identically distributed random elements with respect to P_ν.

Theorem 4.2.3 Let $\{X_{nk}: 1 \leq k \leq n, n \geq 1\}$ be an array of row-wise exchangeable random elements in a separable Banach space that converges in the second mean for each k. If for each $f \in E^*$

$$\rho_n(f) = E[f(X_{n1})f(X_{n2})] \to 0 \text{ as } n \to \infty,$$

then

(i) $E[f(X_{\infty 1})f(X_{\infty 2})] = 0$,

(ii) $E_\nu(X_{\infty 1}) = 0$ for each P_ν with μ_∞-probability one,

and

(iii) $E(X_{\infty 1}) = 0$.

Proof: Since $X_{nk} \overset{2}{\to} X_{\infty k}$ each k, Problem 4.7 shows that

$$\sup_n E||X_{nk}||^2 < \infty \quad \text{and} \quad E||X_{\infty k}||^2 < \infty \qquad (4.2.2)$$

for each k. Using (4.2.2) and convergence in the 2nd mean, for each $f \in E^*$

$$|E[f(X_{n1})f(X_{n2})] - E[f(X_{\infty 1})f(X_{\infty 2})]|$$

$$\leq |E[f(X_{n1})f(X_{n2}) - E[f(X_{\infty 1})f(X_{n2})]|$$

$$+ |E[f(X_{\infty 1})f(X_{n2})] - E[f(X_{\infty 1})f(X_{\infty 2})]|$$

$$\leq E(|f(X_{n1}) - f(X_{\infty 1})||f(X_{n2})|)$$

$$+ E(|f(X_{n2}) - f(X_{\infty 2})||f(X_{\infty 1})|)$$

$$\leq ||f||^2[E(||X_{n1} - X_{\infty 1}||||X_{n2}||)$$

$$+ E(||X_{n2} - X_{\infty 2}||||X_{\infty 1}||)]$$

$$\leq ||f||^2[(E||X_{n1} - X_{\infty 1}||^2)^{\frac{1}{2}}(E||X_{n2}||^2)^{\frac{1}{2}}$$

$$+ (E||X_{n2} - X_{\infty 2}||^2)^{\frac{1}{2}}(E||X_{\infty 1}||^2)^{\frac{1}{2}}]$$

$$\to 0 \text{ as } n \to \infty.$$

Thus,

$$E[f(X_{\infty 1})f(X_{\infty 2})] = 0. \tag{4.2.3}$$

Since convergence in the second mean implies convergence in mean, it follows from Theorem 4.2.2 that the sequence $\{X_{\infty k}: k \geq 1\}$ is exchangeable. It follows from (4.2.1) and (1.3.5) that

$$E[f(X_{\infty 1})f(X_{\infty 2})] = \int_F E_\nu[f(X_{\infty 1})f(X_{\infty 2})]d\mu_\infty(P_\nu)$$

$$= \int_F [E_\nu f(X_{\infty 1})]^2 d\mu_\infty(P_\nu)$$

$$= 0 \ \mu_\infty \text{ a.s. for each } f \in E^*.$$

Thus, $[E_\nu f(X_{\infty 1})]^2 = 0 \ \mu_\infty$ a.s. which implies that $E_\nu f(X_{\infty 1}) = 0 \ \mu_\infty$ a.s. for each $f \in E^*$. From the definition of a Pettis integral, $f(E_\nu(X_{\infty 1})) = E_\nu(f(X_{\infty 1})) = 0 \ \mu_\infty$ a.s. for each $f \in E^*$ which implies that $E_\nu(X_{\infty 1}) = 0 \ \mu_\infty$ a.s. (see Problem 4.8). Also, $EX_{\infty 1} = \int_F E_\nu(X_{\infty 1}) d\mu_\infty(P_\nu) = 0$. ///

4.3 CONDITIONAL EXPECTATIONS AND MARTINGALE TECHNIQUES

A few properties of conditional expectations and martingales in Banach spaces will be developed in this section. Space does not permit an extensive development of these concepts. Thus, attention will be narrowly restricted to the properties which are needed in Chapters 5 and 6. Scalora (1961) and Diestel and Uhl (1977) provide very detailed development of these concepts.

For a random element X such that $E||X|| < \infty$ and a sub-σ-field $C \subset A$, then the conditional expectation, $E(X|C)$, of X given C can be defined (uniquely, with probability one) to be the C-measurable random element such that $E[E(X|C)I_A] = E[XI_A]$ for each $A \in C$. [Diestel and Uhl (1977)]. Moreover, the usual properties for conditional expectations can be shown to hold for random elements. In particular, Property 1.2.4

$$\frac{1}{m} \Sigma_{k=1}^m X_k = E[X_1|F_m] \quad \text{a.s.} \tag{4.3.1}$$

holds for exchangeable random elements where F_m is the permutable σ-field of order m or is generated by $\{\Sigma_{k=1}^m X_k, X_{m+1}, \ldots\}$. Problem 4.9 shows that $\{E[X_1|F_m]: m \geq 1\}$ is a reverse martingale. But, the martingale convergence theorem need not hold for all separable Banach spaces (see Chatterji (1976) for a specific example). In fact, the existence of a limit random element in the martingale convergence theorem is related to the Banach space having the Radon-Nikodym property (rf. Diestel and Uhl (1977)).

However, the following inequality for reverse martingales is available for all Banach spaces and in combination with (4.3.1) is very useful in obtaining the almost sure convergence of $\frac{1}{m}\Sigma_{k=1}^{m} X_k$.

Lemma 4.3.1 If $\{X_n : n \geq 1\}$ is a reverse martingale with respect to $\{F_n\}$, then for any $\varepsilon > 0$

$$P[\sup_{n\geq m}||X_n|| > \varepsilon] \leq \frac{1}{\varepsilon} \int_{[\sup_{n\geq m}||X_n|| > \varepsilon]} ||X_1||dP. \quad (4.3.2)$$

Proof: Since $\{X_n\}$ is a reverse martingale, $X_{n+m} = E(X_n|F_{n+m})$ for all n, m positive integers. Thus, $||X_{n+m}|| = ||E(X_n|F_{n+m})|| \leq E(||X_n|||F_{n+m})$ a.s., and $\{||X_n||: n \geq m\}$ is a nonnegative, reverse submartingale sequence of random variables. The proof then follows from Tucker (1967) p. 224 with only slight modifications. ///

A sequence of random elements $\{X_n\}$ is said to be a martingale difference sequence with respect to the sequence of sub-σ-fields $\{F_n\}$ if $E(X_n|F_{n-1}) = 0$ a.s. for each n. Martingale difference sequences are useful in obtaining results for sequences of dependent random elements. However, a crucial pth moment inequality (similar to (4.1.2)) holds in general if and only if the Banach space is p-uniformly smoothable (see Section 4.1).

The last two lemmas in this section are used in Chapter 6, and their proofs illustrate some of the techniques for conditional expectations of random elements.

Lemma 4.3.2 Let Y be a random element such that $E||Y|| < \infty$ and let $\mathcal{C}$ be a sub-σ-field of $\mathcal{A}$. Then if f is a continuous, linear function of E_1 into E_2 (both Banach spaces),

$$f(E(Y|\mathcal{C})) = E(f(Y)|\mathcal{C}) \text{ a.s.} \qquad (4.3.3)$$

Proof: Since both sides of (4.3.3) are $\mathcal{C}$-measurable, it suffices to show that $\int_B f(E(Y|\mathcal{C}))dP = \int_B E(f(Y)|\mathcal{C})dP$ for all $B \in \mathcal{C}$. Hence, for $B \in \mathcal{C}$

$$\begin{aligned}\int_B E(f(Y)|\mathcal{C})dP &= \int_B f(Y)dP \\ &= \int_\Omega f(Y)I_B dP \\ &= \int_\Omega f(YI_B)dP \\ &= E(f(YI_B)). \qquad (4.3.4)\end{aligned}$$

Using the definitions of the conditional expectation and Theorem 2.3.7 (iv) of Taylor (1978)

$$\begin{aligned}E(f(YI_B)) &= f(E(YI_B)) \\ &= f(\int_\Omega YI_B\, dP) \\ &= f(\int_B Y\, dP). \qquad (4.3.5)\end{aligned}$$

Similar to (4.3.5),

$$
\begin{aligned}
f(\int_B Y\, dP) &= f(\int_B E(Y|C)dP) \\
&= f(\int_\Omega E(Y|C) I_B dP) \\
&= f(E[E(Y|C) I_B]). \\
&= E(f[E(Y|C)] I_B) \\
&= \int_\Omega f(E(Y|C)) I_B dP \\
&= \int_B f(E(Y|C))dP. \qquad (4.3.6)
\end{aligned}
$$

Combining (4.3.4), (4.3.5), and (4.3.6)

$$\int_B f(E(Y|C))dP = \int_B E(f(Y)|C)dP$$

for all $B \in C$. Therefore,

$$f(E(Y|C)) = E(f(Y)|C) \quad \text{a.s.} \qquad ///$$

<u>Lemma 4.3.3</u> Let $\{Y_n : n \geq 1\}$ be a reverse martingale. If f is a continuous, linear function of E_1 into E_2, then $\{f(Y_n): n \geq 1\}$ is a reverse martingale.

<u>Proof</u>: Since $\{Y_n\}$ is a reverse martingale, $E(Y_n | Y_{n+1}, Y_{n+2}, \ldots) = Y_{n+1}$ a.s. For $A \in \sigma\{f(Y_{n+1})\}$,

there exists an $B \in \mathcal{B}(E_2)$ such that $A = (f(Y_{n+1}))^{-1}(B) =$

$(Y_{n+1})^{-1}(f^{-1}(B)) \in \sigma\{Y_{n+1}\}$. Thus, $\sigma\{f(Y_{n+1})\} \subset \sigma\{Y_{n+1}\}$ implies that $\sigma\{f(Y_{n+1}), f(Y_{n+2}), \ldots\} \subset \sigma\{Y_{n+1}, Y_{n+2}, \ldots\}$. From (4.3.3)

$$E(f(Y_n)|f(Y_{n+1}), f(Y_{n+2}), \ldots)$$

$$= E(E(f(Y_n)|Y_{n+1}, Y_{n+2}, \ldots)|f(Y_{n+1}), f(Y_{n+2}), \ldots)$$

$$= E(f(E(Y_n|Y_{n+1}, Y_{n+2}, \ldots))|f(Y_{n+1}), f(Y_{n+2}), \ldots)$$

$$= E(f(Y_{n+1})|f(Y_{n+1}), f(Y_{n+2}), \ldots)$$

$$= f(Y_{n+1}) \text{ a.s.}$$ ///

4.4 PROBLEMS

4.1 Let X be a random element in the space $E = \ell^p$, $p \geq 1$, such that $E||X|| < \infty$. Use the fact that $f_n: X \to R$ by $f_n(x) = x_n$ (where $x = (x_1, x_2, \ldots)$) is in E^* to show that $EX = (EX_1, EX_2, \ldots)$

4.2 A Banach space is said to be <u>uniformly convex</u> if for each $\varepsilon > 0$ there exists $\delta > 0$ such that $||x|| \leq 1$, $||y|| \leq 1$, and $||x+y|| > 2 - 2\delta$ implies that $||x-y|| < \varepsilon$ for all $x, y \in E$. Show that every uniformly convex Banach space is B-convex.

4.3 Give an example of a B-convex space which is not uniformly convex.

4.4 Show that condition (T) reduces to uniform integrability when $E = R$.

4.5 Show that for random elements in a finite-dimensional Banach space, $\sup_n E||X_n||^r < \infty$ for some $r > 0$ implies that $\{X_n\}$ are tight.

4.6 In the proof of Theorem 4.2.1, define $(X_{n1}, \ldots, X_{nk})$: $\Omega \to E\times\ldots\times E$ and show that $X_{ni} \xrightarrow{r} X_{\infty i}$ for each $1 \leq i \leq k$ implies $(X_{n1}, \ldots, X_{nk}) \xrightarrow{r} (X_{\infty 1}, \ldots, X_{\infty k})$.

4.7 In the proof of Theorem 4.2.3, verify (4.2.2).

4.8 In the proof of Theorem 4.2.3, use the separability of E and the Hahn Banach theorem to establish $E_\nu(X_{\infty 1}) = 0$ μ_∞ a.s. when $f(E_\nu(X_{\infty 1})) = 0$ μ_∞ a.s. for each $f \in E^*$.

4.9 Use the fact that $E[X|\mathcal{C}] = E[E[X|\mathcal{B}]|\mathcal{C}] = E[E[X|\mathcal{C}]|\mathcal{B}]$ a.s. when the sub-σ-fields satisfy $\mathcal{C} \subset \mathcal{B} \subset \mathcal{A}$, to show that $\{E[X|\mathcal{F}_m]: m \geq 1\}$ is a reverse martingale where $E||X|| < \infty$ and $\mathcal{F}_m \supset \mathcal{F}_{m+1}$ for all m.

4.10 Show that the distribution of $(X_1, \ldots, X_n)$ is completely determined by the values $E[f(X_1, \ldots, X_n)]$ where f is a bounded, Borel function from E^n into R and $X_1, \ldots, X_n$ are random elements in E.

CHAPTER V

CENTRAL LIMIT THEOREMS FOR EXCHANGEABLE RANDOM ELEMENTS

5.0 INTRODUCTION

In this chapter, central limit theorems for exchangeable random variables are generalized to the case of random elements in a Banach space E. Both approaches of Chapter II carry over: the use of the de Finetti representation and the martingale techniques. The infinite-dimensional nature of the Banach space introduces the additional problem of finding conditions which will yield tightness of sequences of row-sums and satisfactory conditions for this involve introducing hypotheses on the geometry of the Banach space E.

The use of the de Finetti representation is restricted to the case of infinite sequences and arrays with infinite rows. Martingale techniques introduced by Weber (1980) were applied to obtain central limit theorems for weighted sums of exchangeable random elements in a Banach space by Daffer (1984).

5.1 CENTRAL LIMIT THEOREMS USING THE DE FINETTI REPRESENTATION

As indicated in Section 4.2, de Finetti's theorem holds in a separable Banach space because a separable Banach space is Polish (Olshen (1974)). In fact, it holds in the conditional independence form since there always exists a regular version of a conditional probability in a Polish space. These facts allow the central limit theorems for row-wise i.i.d. arrays $\{X_{nk}\}$ of random elements in a separable Banach space E to be used to obtain weak convergence results for row-wise exchangeable arrays in E.

In this chapter, E will always denote a separable Banach space. The following notation will be used: $\mathcal{B}(E)$ is the Borel σ-field on E; $E^\infty = E \times E \times \ldots$ the countable product space; $\mathcal{B}(E^\infty)$ the product σ-field (generated by finite-dimensional cylinders); F the set of probability measures on $(E, \mathcal{B}(E))$; M_E^1 the σ-field on F generated by the mappings $\nu \to \nu(B)$, $B \in \mathcal{B}(E)$. Note that E^∞, with the product topology, is Polish (Hoffmann-Jørgensen (1970), Theorem 17, p. 27) and that $\mathcal{B}(E^\infty)$ is the Borel σ-field of the product topology (Hoffmann-Jørgensen (1970), Proposition 4, p. 49).

Let $\{X_n\}$ be an exchangeable sequence of random elements in E. If we denote by σ the (symmetric) probability measure induced by $\{X_n\}$ on $(E^\infty, \mathcal{B}(E^\infty))$, we have by de Finetti's theorem

$$\sigma = \int_F \nu^\infty \mu(d\nu) \qquad (5.1.1)$$

where μ is a probability measure on (F, M_E^1) (the "mixing measure") and $\nu^\infty = \times_{i \in N} \nu_i$, $\nu_i = \nu$, $\forall i \in N$, is the product probability on $(E^\infty, \mathcal{B}(E^\infty))$ with marginals ν.

<u>Definition 5.1.1</u> A sequence $\{X_n\}$ of random elements in E satisfying $E||X_n||^2 < \infty$, $\forall n \in N$, is said to be <u>weakly uncorrelated</u> if $\mathrm{cov}(f(X_k), f(X_n)) = 0$ whenever $k \neq n$, for each $f \in E^*$. (E^* is the dual of E).

We state the following analogue of Lemma 2.1.1 (for r=1) for a separable Banach space E.

<u>Lemma 5.1.1</u> Let $\{X_n\}$ be a sequence of exchangeable random elements in E satisfying $EX_1 = 0$ and $E||X_1||^2 < \infty$.

Then $\{X_n\}$ are weakly uncorrelated if and only if $E_\nu X_1 = 0$ for μ-almost every $\nu \in F$, where μ is the mixing measure in the de Finetti representation (5.2.1) of the probability measure of $\{X_n\}$ and $E_\nu X_1 = \int_E x_1 \nu(dx_1)$.

Proof: See Theorem 4.2.3.

Theorem 5.1.2 For every sequence $\{X_n\}$ of exchangeable, weakly uncorrelated random elements in E satisfying $EX_1 = 0$, $E||X_1||^2 < \infty$, we have $n^{-\frac{1}{2}} \sum_{k=1}^{k_n} X_k \xrightarrow{W} Z$, a random element in E, if and only if E is type-2 Rademacher.

Proof: Suppose E is type 2. Since $E||X_1||^2 < \infty$, $E_\nu||X_1||^2 = \int_E ||x_1||^2 \nu(dx_1) < \infty$ for μ- almost every $\nu \in F$ (Problem 5.2). Let $Y_n = n^{-\frac{1}{2}}(X_1 + \ldots + X_n)$ and define the probability measures ν_{Y_n}, $\sigma_{Y_n} \in F$ by $\nu_{Y_n}(B) = \nu^\infty\{Y_n \in B\}$ and $\sigma_{Y_n}(B) = \sigma\{Y_n \in B\}$, $\forall B \in \mathcal{B}(E)$. Then from (5.1.1) we get $\sigma_{Y_n}(B) = \int_F \nu_{Y_n}(B)\mu(d\nu), \forall B \in \mathcal{B}(E)$ (see Problem 5.3). Since $\{X_n\}$ are weakly uncorrelated, we have by Lemma 5.1.1 that $E_\nu X_1 = 0$, μ-a.e. Since E is type 2, by the result of Hoffmann-Jørgensen and Pisier (1976), ν_{Y_n} converges weakly to a centered Gaussian measure γ_ν, μ-a.e. By Billingsley (1968), Theorem 2.1, this implies that $\overline{\lim}_{n\to\infty} \nu_{Y_n}(B) \leq \gamma_\nu(B)$, for every closed $B \in \mathcal{B}(E)$. By Fatou's Lemma we have $\overline{\lim}_{n\to\infty} \sigma_{Y_n}(B) \leq \int_F \overline{\lim}_{n\to\infty} \nu_{Y_n}(B)\mu(d\nu) \leq \int_F \gamma_\nu(B)\mu(d\nu)$, for every closed $B \in \mathcal{B}(E)$. Another application of Theorem 2.1 of Billingsley yields weak convergence of $\{Y_n\}$ to a random element Z with probability measure $\int_F \gamma_\nu \mu(d\nu)$.

The converse follows from the result of Hoffmann-Jørgensen and Pisier (1976) by restricting attention to i.i.d. sequences $\{X_n\}$. ///

Remark 5.1.1 As seen in the proof of Theorem 5.1.2, when weak convergence takes place the limit random element Z has probability measure $\int_F \gamma_\nu \mu(d\nu)$, and characteristic functional $\phi(f) = \int_F [\exp(-\frac{1}{2}E_\nu f^2(X_1))]\mu(d\nu) = \int_F [\exp(-\frac{1}{2}\int_E f^2(x)\nu(dx))]\mu(d\nu)$ (see Problem 5.4). We thus get for the limit a "mixture of Gaussians".

The following result illustrates how sufficient conditions for weak convergence of independent random elements can be used, in connection with the de Finetti representation, to obtain similar results for exchangeable random elements.

The following simple sufficient conditions for weak convergence of weighted sums of i.i.d. random elements can be gleaned from Theorem 5.9 of Araujo and Giné (1980) (rf. also de Acosta et al. (1978), Corollary 2.15).

Lemma 5.1.3 Let $\{X_n\}$ be a sequence of i.i.d. random elements in a type-2 Banach space E satisfying $EX_1 = 0$ and $E||X_1||^2 < \infty$. Let $\{a_{nk}\}$, $k = 1, \ldots, k_n$, $k_n \to \infty$, be a sequence of real constants satisfying (i) $\bar{a}_n = \max_{1\le k\le k_n} |a_{nk}| = O(k_n^{-\frac{1}{2}})$ and (ii) $A_n = \sum_{k=1}^{k_n} a_{nk}^2 \to 1$ as $n \to \infty$. Then $S_n = \sum_{k=1}^{k_n} a_{nk}X_k \overset{w}{\to} Z$, a centered Gaussian random element in E, with characteristic functional $\exp(-\frac{1}{2}Ef^2(X_1))$, $f \in E^*$.

Proof: Put $X_{nk} = a_{nk}X_k$. By (i) the array $\{X_{nk}\}$ is infinitesimal. For every $\varepsilon > 0$ we have

$$\sum_{k=1}^{k_n} P[||X_{nk}|| > \varepsilon] \leq k_n P[||X_1|| > \bar{a}_n^{-1}\varepsilon]$$

$$\leq k_n \frac{\bar{a}_n^{-2}}{\varepsilon^2} E||X_1||^2 I_{[||X_1|| > \bar{a}_n^{-1}\varepsilon]} \to 0,$$ by (i). Also, for $f \in E^*$, $\sum_{k=1}^{k_n} Ef^2(X_{nk}) = \sum_{k=1}^{k_n} a_{nk}^2 Ef^2(X_1) \to \sigma_f^2 = Ef^2(X_1)$, by (ii).

Now it is classical that any separable Banach space is isometrically isomorphic to a subspace of $C[0,1]$, and that $C[0,1]$ possesses a Schauder basis. Thus, E can be mapped into $C[0,1]$ and results on probability measures pulled back to E. To avoid unwieldy notation we assume that E possesses a Schauder basis $\{e_n\}$. Thus, $x \in E$ has a unique representation $x = \sum_{n\in N} x_n e_n$, and we have the partial sum and residual operators defined by $P_m x = \sum_{n=1}^{m} x_n e_n$ and $Q_m x = (I - P_m)\, x = \sum_{n=m+1}^{\infty} x_n e_n$. Then $F_m = P_m E = \{P_m x : x \in E\}$ is a finite-dimensional subspace of E. For $\varepsilon > 0$, let $F_m^{\varepsilon} = \{x \in E : \inf_{y\in F_m} ||x-y|| < \varepsilon\}$. Then $x \in F_m^{\varepsilon}$ if $||Q_m x|| < \varepsilon$. We need to show that for every $\varepsilon > 0$, there is m such that

$$\sup_{n\in N} Ed^2(S_n, F_m) < \varepsilon$$

(where $d(x,F_m) = \inf\{||x-y|| : y \in F_m\}$). But

$$\sup_n E(\inf_{y\in F_m} ||S_n - y||^2) = \sup_n E \inf_{y\in F_m} ||Q_m S_n - y||^2$$

$$\leq \sup_n E||Q_m S_n||^2 \leq \sup_n E||\sum_{k=1}^{k_n} Q_m a_{nk} X_k||^2$$

$$\leq \sup_n C \sum_{k=1}^{k_n} E||Q_m a_{nk} X_k||^2 \quad \text{using the type-2 inequality (4.1.2)}$$

$$= C \sup_n A_n \, E||Q_m X_1||^2 \to 0, \text{ as } m \to \infty, \text{ by (ii).} \qquad ///$$

<u>Theorem 5.1.4</u> Let $\{X_n\}$ be a sequence of exchangeable, weakly uncorrelated random elements in a type-2 Banach space E with $EX_1 = 0$ and $E||X_1||^2 < \infty$. Let $\{a_{nk}\}$ $k = 1, \ldots, k_n$, $k_n \to \infty$, be a sequence of real constants satisfying (i) $\bar{a}_n = \max_{1 \leq k \leq k_n} |a_{nk}| = \mathcal{O}(k_n^{-\frac{1}{2}})$ and (ii) $A_n = \sum_{k=1}^{k_n} a_{nk}^2 \to 1$, as $n \to \infty$. Then $S_n = \sum_{k=1}^{k_n} a_{nk} X_k \xrightarrow{w} Z$, a centered random element in E, whose probability measure is a "mixture of Gaussians" $\int_F \gamma_\nu \mu(d\nu)$.

<u>Proof</u>: By (5.1.1), $\sigma = \int_F \nu^\infty \mu(d\nu)$, where σ is the symmetric measure of the sequence $\{X_n\}$. With $\sigma_{S_n}(B) = \sigma\{S_n \in B\}$ and $\nu_{S_n}(B) = \nu^\infty\{S_n \in B\}$, $B \in \mathcal{B}(E)$, we have $\sigma_{S_n} = \int_F \nu_{S_n} \mu(d\nu)$. Now for ν_{S_n} we have $E_\nu X_1 = 0$ for μ-almost every ν, by Lemma 5.1.1 and $E_\nu||X_1||^2 < \infty$. Hence by Lemma 5.1.3, $\nu_{S_n} \xrightarrow{w} \gamma_\nu$, where γ_ν is a centered Gaussian measure on E, with characteristic functional $\exp(-\frac{1}{2} E_\nu f^2(X_1))$, $f \in E^*$. As in the proof of Theorem 5.1.2, we get $\sigma_{S_n} \xrightarrow{w} \int_F \gamma_\nu \mu(d\nu)$. ///

<u>Remark 5.1.2</u> It is apparent that the hypotheses of Lemma 5.1.3, and consequently those of Theorem 5.1.4, can be weakened, but at the expense of introducing further conditions on the sequence $\{X_n\}$. Thus, $\bar{a}_n = \mathcal{O}(k_n^{-\frac{1}{2}})$ could

be replaced by $\bar{a}_n = o(1)$ together with $P[||X_1|| > \bar{a}_n^{-1}\varepsilon] = o(k_n^{-1})$. Also, the requirement that E be type-2 can be dropped if it is assumed that $\lim_{m\to\infty} \sup_n E||Q_m S_n||^2 = 0$ or, say, simply that $\{S_n\}$ is a tight sequence, but here we forfeit conditions in terms of the individual summands, which are much easier to verify.

5.2 CENTRAL LIMIT THEOREMS USING MARTINGALE METHODS

Martingale methods seem particularly useful for obtaining central limit theorems in a (separable) Banach space E for several reasons, in spite of the fact that their application typically requires more stringent geometric conditions on E than were needed in the previous section. This is a consequence of the fact that the validity of the "type-p inequality" (4.1.2) for martingale differences requires E to be p-uniformly smoothable (cf. Section 4.1). However, the classical Banach spaces ℓ^p and L^p are p-uniformly smooth for $1 < p \leq 2$ and 2-uniformly smooth for $2 \leq p < \infty$.

A gratifying feature of martingale methods when considering arrays of row-wise exchangeable random elements is that they avoid cumbersome hypotheses on the asymptotic behaviour of mixing measures of certain sequences of sets of probability measures corresponding to the rows of the array: hypotheses which would be necessary to provide for weak convergence of the integrands in the de Finetti representation. The question of such unpleasant hypotheses is complicated in a Banach space by the necessity of providing some conditions which would be sufficient to ensure tightness of the integrands in the de Finetti representation, at least for sets of measures with sufficiently great probability with respect to the mixing measures. A still more useful aspect of martingale methods is (as was noted in Chapter II) that they involve no mixing measures whatever and are applicable to triangular arrays - arrays with finite rows. Tightness conditions can be formulated simply, in terms of the probability measures of the rows of the array.

We shall need a Banach space version of the following central limit theorem for real-valued martingale difference sequences, which follows from Theorem 3.2 of Hall & Heyde (1980).

<u>Theorem 5.2.1</u> Let $\{X_{nk}, F_{nk}\}$, $k = 1, \ldots, k_n$, $k_n \uparrow \infty$, be a martingale difference array in R satisfying

(i) $\max_{1 \le k \le k_n} |X_{nk}| \xrightarrow{p} 0$; (ii) $\sup_{n \in N} E \max_{1 \le k \le k_n} X_{nk}^2 < \infty$;

(iii) $\sum_{k=1}^{k_n} X_{nk}^2 \xrightarrow{p} \eta^2$, a random variable, a.s. finite and measurable with respect to $F_0 = \bigcap_{n \in N} F_{n0}$. Then $S_n = \sum_{k=1}^{k_n} X_{nk} \xrightarrow{w} Z$, a random variable whose distribution is a mixture of normals, with characteristic function $E \exp(-\frac{1}{2} t^2 \eta^2)$.

<u>Theorem 5.2.2</u> Let $\{X_{nk}, F_{nk}\}$, $k = 1, \ldots, k_n$, $k_n \uparrow \infty$, be a martingale difference array in a Banach space E satisfying

(i) $\max_{1 \le k \le k_n} ||X_{nk}|| \xrightarrow{p} 0$;

(ii) $\sup_{n \in N} E \max_{1 \le k \le k_n} ||X_{nk}||^2 < \infty$;

(iii) $\sum_{k=1}^{k_n} f^2(X_{nk}) \xrightarrow{p} \eta_f^2$, a random variable, a.s. finite and measurable with respect to $F_0 = \bigcap_{n \in N} F_{n0}$, for every $f \in E^*$.

(iv) $\{S_n\}$ is a tight sequence, where $S_n = \sum_{k=1}^{k_n} X_{nk}$.

Then $S_n = \sum_{k=1}^{k_n} X_{nk} \xrightarrow{w} Z$, a random element in E, mixture of Gaussians, with characteristic functional $\phi(f) = E \exp(-\frac{1}{2}\eta_f^2)$, $f \in E^*$.

Proof: Let $f \in E^*$. Then, $\{f(X_{nk}), F_{nk}\}$, $k = 1, \ldots, k_n$, is a martingale difference array in R, and Theorem 5.2.1 yields $f(S_n) = \sum_{k=1}^{k_n} f(X_{nk}) \xrightarrow{w} Z_f$, a random variable with characteristic function $E \exp(-\frac{1}{2}t^2\eta_f^2)$, where $\eta_f^2 = \lim_{n\to\infty} \sum_{k=1}^{k_n} f^2(X_{nk})$, in distribution.

This means that all weak limits of subsequences of $\{S_n\}$ are the same: the probability measure whose marginals are those of the random variables Z_f, with characteristic functions $E \exp(-\frac{1}{2}t^2\eta_f^2)$, $f \in E^*$ (use Billingsley (1968), Theorem 5.1 and the continuity of $f \in E^*$). Now by (iv) $\{S_n\}$ is tight, hence by Prokhorov's theorem (Billingsley (1968)) conditionally compact, and so every subsequence of $\{S_n\}$ contains a further subsequence which converges weakly to this probability measure. By Billingsley (1968), Theorem 2.3, the whole sequence converges weakly to this probability measure, which we denote by μ. For the characteristic functional of γ, we have $\phi(f) = \int_E e^{if(x)}\gamma(dx) = \int_R e^{iu} f\circ\gamma(du) = \phi_f(1)$, where $\phi_f(t) = \int_R e^{itu} f\circ\gamma(du) = E e^{\frac{-t^2\eta_f^2}{2}}$. Thus $\phi(f) = E\, e^{-\frac{1}{2}\eta_f^2}$, $f \in E^*$, as claimed. ///

A central point in proving central limit theorems for random elements in a Banach space is establishing the tightness of row-sums that is required in Theorem 5.2.2. To enable this to be done for arrays of exchangeable random elements in terms of conditions on the individual summands, the following lemma is helpful.

If E possesses a Schauder basis $\{e_n\}$ and if $x = \sum_{n=1}^{\infty} a_n e_n \in E$, then $P_m x = \sum_{n=1}^{m} a_n e_n$ and $Q_m x = (I - P_m)\, x = \sum_{n=m+1}^{\infty} a_n e_n$ defines, for each $m \in N$, the partial sum and residual operators on E, respectively. It follows from the proof of Lemma 4, p. 259, of Dunford & Schwartz (1963), that a set $K \subset E$ is conditionally compact if and only if

(i) $\sup_{x \in K} ||x|| < \infty$; and

(ii) to every $\varepsilon > 0$ there is $m_\varepsilon \in N$ such that for $m \geq m_\varepsilon$

$$\sup_{x \in K} ||Q_m x|| < \varepsilon.$$

<u>Lemma 5.2.3</u> Let $\{X_n\}$ be a sequence of random elements in a Banach space E with a Schauder basis, such that $\{||X_n||^p\}$ is uniformly integrable, where $1 \leq p < \infty$. Then $\{X_n\}$ is tight if and only if

$$\lim_{m \to \infty} \sup_{n \in N} E||Q_m X_n||^p = 0.$$

<u>Proof</u>: Let $\{X_n\}$ be tight and $\varepsilon > 0$ given. Choose α such that $E||X_n||^p I_{[||X_n|| > \alpha]} < \varepsilon/2$, $\forall n \in N$. Next, find K compact, such that $P[X_n \notin K] < \varepsilon/2\alpha^p$, $\forall n \in N$. Then $E||X_n||^p I_{[X_n \notin K]} \leq \alpha^p P[X_n \notin K] + E||X_n||^p I_{[||X_n|| > \alpha]}$ $< \frac{\varepsilon}{2} + \frac{\varepsilon}{2} = \varepsilon$. Now since K is compact, we have

$$\lim_{m \to \infty} \sup_n E||Q_m X_n||^p = \lim_{m \to \infty} \sup_n E||Q_m X_n I_{[X_n \notin K]}||^p + 0$$

$$\leq C \sup_n E||X_n||^p I_{[X_n \notin K]} < C\varepsilon, \text{ where } C = \sup_m ||Q_m||^p.$$

Since ε is arbitrary, $\lim_{m \to \infty} \sup_n E||Q_m X_n||^p = 0$.

Conversely, suppose $\lim_{m \to \infty} \sup_n E||Q_m X_n||^p = 0$. Let $\varepsilon > 0$ be given. To every $k \in N$, there is $m_k \in N$ such that $E||Q_m X_n||^p < k^{-1} 2^{-k-1} \varepsilon$, for all n, and $m \geq m_k$. By

$P[||X_n||^p > \alpha] < \alpha^{-1}E||X_n||^p I_{[||X_n|| > \alpha]}$, find α such that $P[||X_n||^p > \alpha] < \varepsilon/2$, for all n. Now let $A_0 = \{x \in E: ||x||^p \leq \alpha\}$ and $A_k = \{x \in E: ||Q_{m_k} x||^p \leq k^{-1}\}$, $k \in N$. Put $K = \bigcap_{k=0}^{\infty} A_k$. It is easily checked that K is conditionally compact. Now $P[X_n \notin K] = P[X_n \in \bigcup_{k=0}^{\infty} A_k^c]$

$$= P[\bigcup_{k=0}^{\infty} [X_n \notin A_k]] \leq P[X_n \notin A_0] + \sum_{k=1}^{\infty} P[X_n \notin A_k]$$

$$= P[||X_n||^p > \alpha] + \sum_{k=1}^{\infty} P[||Q_{m_n} X_n||^p > k^{-1}] < \frac{\varepsilon}{2}$$

$$+ \sum_{k=1}^{\infty} k\, E||Q_{m_k} X_n||^p < \frac{\varepsilon}{2} + \varepsilon \sum_{k=1}^{\infty} 2^{-k-1} = \varepsilon.$$ Since ε is arbitrary, $\{X_n\}$ is tight. ///

The instrument which allows for weak convergence results to be proved in terms of conditions on the individual summands in the row-sums of a sequence or array is the "type-p inequality". As we shall need this for sums of martingale differences, some smoothness assumptions about the space E will need to be made.

The following fact due to Hoffmann-Jørgensen and Pisier (1976) will be needed: If E is p-uniformly smoothable then $E||\sum_{k=1}^{n} X_k||^p \leq C \sum_{k=1}^{n} E||X_k||^p$ for every martingale difference sequence $X_1, \ldots, X_n$ with $E||X_k||^p < \infty$, $k = 1, \ldots, n$, where C is independent of n.

We now prove a central limit theorem for triangular arrays of row-wise exchangeable random elements.

For $\{X_{nk}\}$, $k = 1, \ldots, k_n$, let $F_{nk} = \sigma\{X_{n1}, \ldots, X_{nk}, \sum_{j=k+1}^{k_n} X_j\}$, $k = 0, 1, \ldots, k_n - 1$; $F_0 = \bigcap_{n=1}^{\infty} F_{n0}$.

Theorem 5.2.4 Let $\{X_{nk}\}$, $k = 1, \ldots, k_n$, $k_n \uparrow \infty$, be an array of row-wise exchangeable random elements in a 2-uniformly smoothable Banach space E. Let $\{a_{nk}\}$, $k = 1, \ldots, k_n$, be an array of real constants. Put $\bar{a}_n = \max_{1 \le k \le k_n} |a_{nk}|$ and $A_n = \sum_{k=1}^{k_n} a_{nk}^2$. Assume:

1° $\bar{a}_n \to 0$ and $\sup_n A_n < \infty$;

2° $\max_{1 \le k \le k_n} ||X_{nk}|| = o(\bar{a}_n^{-1})$;

3° $Ef(X_{n1})f(X_{n2}) \to 0$, $\forall f \in E^*$;

4° $\{X_{n1}\}$ is a tight sequence;

5° $\{||X_{n1}||^2\}$ is uniformly integrable;

6° $\sum_{k=1}^{k_n} a_{nk}^2 f^2(X_{nk}) \xrightarrow{P} \eta_f^2$, a random variable, a.s. finite, and F_0-measurable, $\forall f \in E^*$.

Then $\sum_{k=1}^{k_n} a_{nk}(X_{nk} - E(X_{nk}|F_{n,k-1})) \xrightarrow{w} Z$, a random element in E whose probability measure is a mixture of Gaussians, with characteristic functional $\phi(f) = E\exp(-\frac{1}{2}\eta_f^2)$, $f \in E^*$.

Proof: Put $Y_{nk} = a_{nk}(X_{nk} - E(X_{nk}|F_{n,k-1}))$. Then $\{Y_{nk}, F_{nk}\}$ is a martingale difference array in E. We must verify the hypotheses of Theorem 5.2.2 for $\{Y_{nk}, F_{nk}\}$.

The mechanics of verifying hypotheses (i), (ii) and (iii) of Theorem 5.2.2 are the same as were involved in verifying conditions (i), (ii) and (iii) of Theorem 2.2.1, and it would be repetitious to reproduce them here.

One comment is in order regarding the proof of (iii). The calculation analogous to (2.2.6) in the proof of Theorem 2.2.2 leads in the present case to the expression (the sum is up to k_n; there is no m_n):

$$\sum_{k=1}^{k_n} a_{nk}^2 |E\, f(X_{nk}) \frac{\sum_{j=k}^{k_n} f(X_{nj})}{k_n-k+1}$$

$$\leq ||f||^2 \sup_n E||X_{n1}||^2 \sum_{k=1}^{k_n} \frac{a_{nk}^2}{k_n-k+1} + A_n |Ef(X_{n1})f(X_{n2})|.$$

This will go to zero if $\sum_{k=1}^{k_n} \frac{a_{nk}^2}{k_n-k+1} \to 0$. This was shown in Problem 2.2.

The new aspect of this proof is the demonstration of (iv) of Theorem 5.2.2: that $\{\sum_{k=1}^{k_n} Y_{nk}\}$ is a tight sequence. As in the proof of Lemma 5.1.3, we assume without loss of generality that E possesses a Schauder basis. Using the linearity of the residual operator Q_m and the 2-uniform smoothability of E, we have

$$E||\sum_{k=1}^{k_n} Q_m Y_{nk}||^2 \leq C\, \Sigma_k\, E||Q_m Y_{nk}||^2$$

$$\leq 2\,C \sum_k a_{nk}^2\, E[\,||Q_m X_{nk}||^2 + ||Q_m E(X_{nk}|F_{n,k-1})||^2]$$

$$\leq 2\,C \sum_k a_{nk}^2\, [E||Q_m X_{nk}||^2 + E(E||Q_m X_{nk}||^2|F_{n,k-1})]$$

$$\leq 4\, CA_n\, E||Q_m X_{n1}||^2 \to 0$$ using 1^0, and using 4^0 and 5^0 in Lemma 5.2.3 (with $p=2$). Thus, $E||\sum_{k=1}^{k_n} Q_m Y_{nk}||$ $\leq [E||\sum_{k=1}^{k_n} Q_m Y_{nk}||^2]^{\frac{1}{2}} \to 0$. But $\{\sum_k Y_{nk}\}$ is uniformly integrable; this follows from

$$\sup_n E||\sum_k Y_{nk}||^2 \leq 4\,C \sup_n A_n \sup_n E||X_{n1}||^2 < \infty.$$

Thus Lemma 5.2.3, applied to $\{\sum_k Y_{nk}\}$, this time for $p = 1$, yields the required tightness. ///

Remark 5.2.1 If the hypothesis of 2-uniform smoothability is reduced to that of p-uniform smoothability with $1 \leq p < 2$, then the method of proof used in Theorem 5.2.4 would require the assumption that $\sup_{n\in N} \sum_{k=1}^{k_n} |a_{nk}|^p < \infty$. But this, together with $\bar{a}_n \to 0$, implies that $A_n \to 0$, by $\sum_{k=1}^{k_n} a_{nk}^2 \leq \bar{a}_n^{2-p} \sum_{k=1}^{k_n} |a_{nk}|^p \to 0$; and $A_n \to 0$ implies that the convergence is degenerate. In fact $\eta_f^2 = 0$, a.s. for every $f \in E^*$, as is seen by

$$P[\sum_{k=1}^{k_n} a_{nk}^2 f^2(X_{nk}) > \varepsilon] < \varepsilon^{-1} A_n ||f||^2 E||X_{n1}||^2 \to 0.$$

A central limit theorem for weighted sums of an infinite sequence $\{X_n\}$ of exchangeable random elements in E can be obtained from Theorem 5.2.4 as follows. Suppose the array $\{X_{nk}\}$ in Theorem 5.2.4 is doubly infinite: consisting of infinite exchangeable rows. Let the σ-fields be augmented as follows: let

$$F_{nk} = \sigma\{X_{n1}, \ldots, X_{nk}, \sum_{j=k+1}^{k_n} X_{nj}, X_{n,k_n+1}, X_{n,k_n+2}, \ldots\}$$

and $F_0 = \bigcap_{n=1}^{\infty} F_{n0}$. Then the proof of Theorem 5.2.4 goes through exactly as before; the only difference in the conclusion is that the centering conditional expectations $E(X_{nk}|F_{n,k-1})$ are measurable with respect to richer σ-fields than in the original statement of the theorem. This is not necessarily a desirable thing in Theorem 5.2.4 but these observations now enable us to formulate the following corollary by specialization.

For a sequence $\{X_n\}$ define the σ-fields

$$\mathcal{F}_{nk} = \sigma\{X_1, \ldots, X_k, \sum_{j=k+1}^{k_n} X_j, X_{k_n+1}, X_{k_n+2}, \ldots\} \text{ and}$$

$\mathcal{F}_0 = \bigcap_{n=1}^{\infty} \mathcal{F}_{n0}$. (Note that if $k_n \uparrow \infty$, $\mathcal{F}_0$ is the exchangeable σ-field of the sequence $\{X_n\}$.) A_n and $\bar{a}_n$ have the same meaning as before.

<u>Corollary 5.2.5</u> Let $\{X_n\}$ be a sequence of exchangeable random elements in a 2-uniformly smoothable Banach space E satisfying (i) $E||X_1||^2 < \infty$ and (ii) $Ef(X_1)f(X_2) = 0$, $\forall f \in E^*$. Let $\{a_{nk}\}$, $k = 1, \ldots, k_n$, $k_n \uparrow \infty$, be an array of real constants satisfying

(a) $\sup_n A_n < \infty$;

(b) $\max_{1 \le k \le k_n} ||X_k|| = o_p(\bar{a}_n^{-1})$;

(c) $\sum_{k=1}^{k_n} a_{nk}^2 f^2(X_k) \xrightarrow{P} \eta_f^2$, a random variable, a.s. finite, $\forall f \in E^*$.

Then $\sum_{k=1}^{k_n} a_{nk}(X_k - E(X_k|\mathcal{F}_{n,k-1})) \xrightarrow{w} Z$, a random element in E whose probability measure is a mixture of Gaussians, with characteristic functional $\phi(f) = E \exp(-\frac{1}{2} \eta_f^2)$, $f \in E^*$.

<u>Proof:</u> Follows from specialization of Theorem 5.2.4. A couple of points need to be made. $\bar{a}_n \to 0$ follows here from (b), except for the case where $X_k = 0$, a.s., in which case the corollary is trivial. 4° and 5° of Theorem 5.2.4

do not apply. Note that from $\bar{a}_n \to 0$ it follows that η_f^2 is measurable with respect to the tail σ-field, hence measurable with respect to F_0, so that this need not be explicitly assumed. ///

Remark 5.2.2 The condition $\sup_n A_n < \infty$ of Corollary 5.2.5 is actually superfluous. To see how to obtain (a) from the other hypotheses, see Daffer (1984) (the argument there can be generalized to hold for the random limit in (c)). This same reference contains a version of Theorem 5.2.4 in which $\bar{a}_n \to 0$ is not assumed, but where the convergence in 6° of Theorem 5.2.4 is to a constant rather than a random variable.

Remark 5.2.3 It can be shown that 5° and 6° of Theorem 5.2.4, together with $\sup_n A_n < \infty$, are sufficient to imply $\sum_{k=1}^{k_n} a_{nk}^2 f^2(X_{nk}) \xrightarrow{L^1} \eta_f^2$, so that the convergence in 6^0 of Theorem 5.2.4 could have been stated as L^1 convergence, without loss of generality. (See Problem 5.5)

It is not clear how to eliminate the centering at conditional expectations in the conclusion of the theorem in the infinite-dimensional case. It is possible to extend a result of N. C. Weber (1980) which is obtained using reverse martingales to the case of random elements in a Banach space E. Centering at conditional expectations is still required but the σ-fields used in the centering random elements are different from those in Theorem 5.2.4. This method, however, works only for the special classical weights $a_{nk} = k_n^{-\frac{1}{2}}$, $k = 1, \ldots, k_n$.

We state this result without proof. The proof follows closely to that of Theorem 2 of Weber (1980), and the proof of tightness of row-sums of a martingale difference array that is needed in the infinite-dimensional case is achieved with the techniques which were used for this purpose in the proof of Theorem 5.2.4. The Banach space E is assumed without loss of generality to possess a Schauder basis and uniform integrability of $\{||X_{n1}||^2\}$ and 2-uniform smoothability of E are used in conjunction with the tightness condition provided by Lemma 5.2.3.

For an array $\{X_{nk}\}$, $k = 1, \ldots, m_n$, define the σ-fields

$S_{nk} = \sigma\{(X_{n1}^{(1)}, \ldots, X_{nk}^{(k)}), X_{n,k+1}, X_{n,k+2}, \ldots, X_{nm_n}\}$,

$k = 1, \ldots, m_n$, where $(X_{n1}^{(1)}, \ldots, X_{nk}^{(k)})$ are the order statistics of $(X_{n1}, \ldots, X_{nk})$.

<u>Theorem 5.2.6</u> Let $\{X_{nk}\}$, $k = 1, \ldots, m_n$, $m_n \uparrow \infty$, be an array of zero-mean, row-wise exchangeable random elements in a 2-uniformly smoothable Banach space E, satisfying $\sup_n E||X_{n1}||^2 < \infty$. Let $\{k_n\}$ be a sequence of integers with $1 \leq k_n \leq m_n$ and $\lim_{n\to\infty} k_n m_n^{-1} = \alpha$, where $0 < \alpha < 1$. Suppose further that

(a) $\lim_{n\to\infty} Ef(X_{n1})f(X_{n2}) = 0$, $\forall f \in E^*$;

(b) $\max_{1\leq k\leq m_n} ||X_{nk}|| = o_p(\sqrt{k_n})$, as $n \to \infty$;

(c) $\{||X_{n1}||^2\}$ is uniformly integrable;

(d) $k_n^{-1} \sum_{k=1}^{k_n} f^2(X_{nk}) \xrightarrow{L^1} \eta_f^2$, a random variable, a.s. finite, $\forall f \in E^*$;

(e) $\{X_{n1}\}$ is a tight sequence in E.

Assume further that either

(i) the σ-fields are nested as follows:

$$\mathcal{S}_{n+1,k} \supset \mathcal{S}_{nk}, \quad \forall n \in \mathbb{N},\ k = n, n+1, \ldots, m_n;$$

or

(ii) η_f^2 is measurable $\bigcap_{n \in \mathbb{N}} \mathcal{S}_{nm_n}$, $\forall f \in E^*$.

Then

$$k_n^{-\frac{1}{2}} \sum_{k=1}^{k_n} (X_{nk} - E(X_{n1} | \mathcal{S}_{nm_n})) \xrightarrow{w} Z, \text{ a centered random}$$

element in E whose probability measure is a mixture of Gaussians, with characteristic functional $E \exp(-\frac{1}{2}(1-\alpha)\eta_f^2)$, $f \in E^*$.

5.3 PROBLEMS

5.1 Prove Lemma 5.1.1.

5.2 Prove the assertion that $E_\nu ||X_1||^2 < \infty$, μ-a.e. (Justify the formula $\int_{E^\infty} ||x_1||^2 \sigma(dx) = \int_{M_E^1} \int_E ||x_1||^2 \nu(dx_1)$, where $X = (X_1, X_2, \ldots) \in E^\infty$.)

5.3 Demonstrate that the map $\nu \to \nu_{Y_n}(B)$, (for fixed $B \in \mathcal{B}_E$) of M_E^1 into R, is measurable. (ν_{Y_n} is defined in the proof of Theorem 5.1.2.)

5.4 Justify the interchange of order of integration:
$\phi(f) = \int_E e^{if(x)} [\int_{M_E^1} \gamma_\nu(dx)\mu(d\nu)] = \int_{M_E^1} [\int_E e^{if(x)} \gamma_\nu(dx)]\mu(d\nu)$.

5.5 Show that 5° and 6° of Theorem 5.2.4 together with $A = \sup_n A_n < \infty$ imply that $\sum_{k=1}^{k_n} a_{nk}^2 f^2(X_{nk}) \xrightarrow{L^1} \eta_f^2$. [Using exchangeability, write

$$E \sum_k a_{nk}^2 f^2(X_{nk})\, I_{[\Sigma a_{nk}^2 f^2(X_{nk}) > \alpha]}$$

$$\leq A||f||^2 \{E||X_{n1}||^2\, I_{[\Sigma a_{nk}^2 f^2(X_{nk}) > \alpha]} I_{[||X_{n1}||^2 \leq \alpha]}$$

$$+ E||X_{n1}||^2 \, I_{[\Sigma a_{nk}^2 f^2(X_{nk}) > \alpha]} I_{[||X_{n1}||^2 > \alpha]}\}.$$

Work with this, to get uniform integrability. $E\eta_f^2 < \infty$ will need to be shown.]

CHAPTER VI

STOCHASTIC CONVERGENCE OF WEIGHTED SUMS OF EXCHANGEABLE RANDOM ELEMENTS

6.0 INTRODUCTION

In this chapter, convergence results for exchangeable random elements in separable Banach spaces will be presented. An exhaustive presentation of these results will not be presented; but, sufficient information is presented to provide an overview of the area and references to results will indicate where additional details are available. Proofs of several of these results will be included to illustrate some of the techniques used in obtaining convergence results. In addition, examples that illustrate some of the strong convergence results will be given.

6.1 WEAK CONVERGENCE RESULTS FOR INFINITE ARRAYS OF EXCHANGEABLE RANDOM ELEMENTS

In this section, weak convergence results will be obtained for weighted sums of exchangeable random elements. The first result by Taylor (1982a) is a weak convergence result for an infinite array of random elements that are row-wise exchangeable. In particular, a weak convergence result will be obtained for random elements in type p Banach spaces, $1 \leq p \leq 2$. Recall from Section 4.2 that

$$P(B_n) = \int_{\mathcal{F}} P_\nu(B_n)\,d\mu_n(P_\nu) \qquad (6.1.1)$$

where n refers to the nth row of some infinite array.

For $\eta > 0$ and an array of constants $\{a_{nk}\}$, define

$$\mathcal{F}(\{a_{nk}\},\eta) = \{P_\nu : \max_{1\leq k\leq n} \|E_\nu(X_{nk} I_{[\|X_{nk}\| \leq |a_{nk}|^{-1}]})\| < \frac{\eta}{\sum_{k=1}^{n} |a_{nk}|}\} \qquad (6.1.2)$$

for each n, where $a_{nk}^{-1} = \infty$ when $a_{nk} = 0$. The measurability of $F(\{a_{nk}\},\eta)$ was established in Chapter 4. Also, let $P(t) = \sup_n P[||X_{n1}||>t]$ for each $t \geq 0$.

Theorem 6.1.1 Let $\{X_{nk}\}$ be random elements in a Banach space of type $p + \delta$, $1 \leq p \leq 2$, for some $\delta > 0$. For each n, let $\{X_{nk}: k \geq 1\}$ be exchangeable. Let $\{a_{nk}\}$ be constants such that

$$\sum_{k=1}^{n} |a_{nk}|^p \leq 1 \text{ and } \max_{1\leq k\leq n} |a_{nk}| \to 0. \tag{6.1.3}$$

If

$$t^p P(t) \to 0 \text{ as } t \to \infty \text{ and}$$

$$\mu_n(F(\{a_{nk}\},\eta)) \to 1 \text{ for each } \eta > 0, \tag{6.1.4}$$

then

$$||\sum_{k=1}^{n} a_{nk}X_{nk}|| \to 0 \text{ in probability.}$$

Proof: Define $Y_{nk} = X_{nk} I_{[||X_{nk}|| \leq |a_{nk}|^{-1}]}$. If $a_{nk} = 0$, then let $|a_{nk}|^{-1} = \infty$ with $t^p P[||X_{n1}|| > t] = 0$ for $t = \infty$. For each n

$$P[\sum_{k=1}^{n} a_{nk}Y_{nk} \neq \sum_{k=1}^{n} a_{nk}X_{nk}] \leq \sum_{k=1}^{n} P[||X_{nk}|| > |a_{nk}|^{-1}]$$

$$\leq \sum_{k=1}^{n} |a_{nk}|^p (|a_{nk}|^{-p} P[||X_{n1}|| > |a_{nk}|^{-1}])$$

which goes to 0 since $\sum_{k=1}^{n} |a_{nk}|^p \leq 1$, $\max_{1 \leq k \leq n} |a_{nk}| \to 0$, and $\sup_n x^p P[||X_{n1}|| > x] \to 0$ as $x \to \infty$. Hence for $\varepsilon > 0$, it suffices to show that $||\sum_{k=1}^{n} a_{nk} Y_{nk}|| \to 0$ in probability. For each n and each T > 0,

$$\int_0^T x^{p+\delta} dP[||X_{n1}|| \leq x] = T^{p+\delta} P[||X_{n1}|| \leq T]$$

$$- (p + \delta) \int_0^T x^{p+\delta-1} P[||X_{n1}|| \leq x]dx$$

$$\leq (p + \delta) \int_0^T x^{p+\delta-1} P[||X_{n1}|| > x]dx. \quad (6.1.5)$$

Given $\eta > 0$, B can be chosen so that $\sup_n P[||X_{n1}|| > x] \leq \eta\delta x^{-p}$ for all $x \geq B$. Then for all n

$$T^{-\delta} \int_0^T x^{p+\delta-1} P[||X_{n1}|| > x]dx$$

$$\leq T^{-\delta} \left(\int_0^B x^{p+\delta-1} P[||X_{n1}|| > x]dx + \eta \int_B^T \delta x^{\delta-1} dx\right)$$

$$\leq T^{-\delta}((p + \delta)^{-1} B^{p+\delta} + \eta\, T^{\delta}) \quad (6.1.6)$$

for T sufficiently large. Letting $F_\varepsilon = F(\{a_{nk}\}, \frac{\varepsilon}{2})$,

$$P[||\sum_{k=1}^{n} a_{nk} Y_{nk}|| > \varepsilon]$$

$$\leq \int_{F_\varepsilon} P_\nu[||\sum_{k=1}^{n} a_{nk}(Y_{nk} - E_\nu Y_{nk})|| > \frac{\varepsilon}{2}] d\mu_n(P_\nu)$$

$$+ \mu_n(F_\varepsilon^c) \quad (6.1.7)$$

for each n. Using (6.1.5) and the moment inequality of a type p + δ Banach space

$$\int_{F_\varepsilon} P_\nu[||\sum_{k=1}^{n} a_{nk}(Y_{nk} - E_\nu Y_{nk})|| > \tfrac{\varepsilon}{2}]d\mu(P\nu)$$

$$\leq K \int_{F_\varepsilon} \sum_{k=1}^{n} E_\nu||a_{nk}(Y_{nk} - EY_{nk})||^{p+\delta} d\mu_n(P_\nu)$$

$$\leq K \int_{F_\varepsilon} \sum_{k=1}^{n} |a_{nk}|^{p+\delta} \int_0^{|a_{nk}|^{-1}} x^{p+\delta} dP_\nu[||X_{nk}|| > x]d\mu_n(P_\nu)$$

$$\leq K \sum_{k=1}^{n} |a_{nk}|^{p+\delta}(p+\delta) \int_{F_\varepsilon}\int_0^{|a_{nk}|^{-1}} x^{p+\delta-1} P_\nu[||X_{n1}||> x]dxd\mu_n(P_\nu)$$

$$\leq K \sum_{k=1}^{n} |a_{nk}|^{p}(|a_{nk}|^{\delta} \int_0^{|a_{nk}|^{-1}} x^{p+\delta-1} P[||X_{n1}|| > x]dx) \tag{6.1.8}$$

for each n where $K = (4/\varepsilon)^{p+\delta}C$ and where the last inequality follows from Tonelli's theorem and $\mu_n(F_\varepsilon) \leq 1$. Thus, the result follows from (6.1.3), 6.1.4), 6.1.6) and 6.1.7). ///

For random variables, (1.3.8) in Section 1.3 and Problem 1.8 related $\mu_n(F(\{a_{nk}\},\eta) \to 1$ to $\rho_n \to 0$. In particular for random variables when $a_{nk} = n^{-1/p}$, $1 \leq p < 2$, then a Marcinkiewicz-Zygmund weak law follows when $(n^{2-2/p})\rho_n \to 0$, that is, the rate that a measure of nonorthogonality must go to zero is inversely related to the size of the weights. Also, when the random elements are uniformly bounded, the set $F(\{a_{nk}\},\eta)$ in (6.1.2) can be defined in terms of the conditional means rather than the truncated conditional means which is indicated by Problem 1.8.

When the random variables $\{||X_{n1}||^p\colon n \geq 1\}$ are uniformly integrable, (6.1.2) can be expressed as

$$F(\{a_{nk}\},\eta) = \{P_\nu\colon ||E_\nu(X_{n1})|| < \eta/\sum_{k=1}^{n}|a_{nk}|\} \ (= F'_\eta)$$

This can be seen by considering the proof of Theorem 6.1.1 and noting that for $\eta = \varepsilon/2$

$$P[||\sum_{k=1}^{n} a_{nk}X_{nk}|| > \varepsilon]$$

$$\leq \int_{F_n'} P_\nu[||\sum_{k=1}^{n} a_{nk}(X_{nk} - EX_{n1})|| > \frac{\varepsilon}{2}]d\mu_n(P_\nu) + \mu_n(F_n'^{\,c})$$

$$\leq \int_{F_n'} P_\nu[||\sum_{k=1}^{n} a_{nk}(Y_{nk} - EY_{nk})|| > \frac{\varepsilon}{4}]d\mu_n(P_\nu) + \mu_n(F_n'^{\,c})$$

$$+ (\frac{4}{\varepsilon})\, C\, 2^{p+1} \sum_{k=1}^{n} |a_{nk}|^p E||X_{n1} I_{[||X_{n1}|| > |a_{nk}|^{-1}]}||^p$$

where the constant C depends only on the $p + \delta$ geometry condition of the Banach space.

It is clear that the condition $x^p \sup_n P[||X_{n1}|| > x] \to 0$ as $x \to \infty$ can be replaced by the condition that $b_n^{-p} P[||X_{n1}|| > b_n^{-1}] \to 0$ as $n \to \infty$ where $b_n = \max_{1 \leq k \leq n} |a_{nk}|$. Mandrekar and Zinn (1980) have shown that when $\{X_n\}$ are independent, identically distributed random elements in a separable Banach space E and when $n^p P[||X_1|| > n] \to 0$ is sufficient for the weak law of large numbers, then E must be of type $p + \delta$ for some $\delta > 0$. Thus, type $p + \delta$ is necessary in Theorem 6.1.1. A very important use of Theorem 6.1.1 occurs when $E = R^m$ since R^m is of type $p + \delta$ for each p, $1 \leq p < 2$, and $p + \delta \leq 2$. Note that $p < 2$ which is expected since under more restrictive conditions than those of Theorem 6.1.1, it follows that $n^{-\frac{1}{2}} \sum_{k=1}^{n} X_{nk}$ converges in distribution to a non-degenerate normal random variable. For spaces which are not of type $p + \delta$, that is, type 1 spaces, a tightness condition must be imposed on the random elements (see Taylor (1978) for a general discussion and examples). A very general condition involving tightness of distributions and moments is given in Definition 4.1.2. Recall that Condition (T) for an array $\{X_{nk}\}$ of random elements is given by

(T): Given $\varepsilon > 0$ there exists a compact subset K such that

$$\sup_{n,k} E||X_{nk} I_{[X_{nk} \notin K]}|| < \varepsilon$$

Using Condition (T) and the continuous linear functionals, a version of Theorem 3.1.3 can be obtained for all separable Banach spaces.

<u>Theorem 6.1.2</u> Let $\{X_{nk}\}$ be random elements in a separable Banach space E such that Condition (T) holds. For each $f \in E^*$ let $\rho_n(f) = \max_{k \neq \ell} |E[f(X_{nk})f(X_{n\ell})]| \to 0.$

(6.1.9)

If $\{a_{nk}\}$ are constants such that

$$\sum_{k=1}^{n} |a_{nk}| \leq 1 \text{ and } \sum_{k=1}^{n} a_{nk}^2 E[f^2(X_{nk})] \to 0 \text{ for each } f \in E^*,$$

then

$$||\sum_{k=1}^{n} a_{nk}X_{nk}|| \to 0 \text{ in probability.}$$

<u>Proof</u>: It can be assumed that E has a Schauder basis $\{b_i\}$ and coordinate functionals $\{f_i\}$ and that $||U_t|| = 1$ and $||Q_t|| = 1$ for all t where $U_t(x) = \sum_{i=1}^{t} f_i(x)b_i$ and $Q_t(x) = x - U_t(x)$ since E can be isometrically embedded in C[0,1]. Given $\varepsilon > 0$ and $\delta > 0$ pick K compact such that

$$\sup_{n,k} E||X_{nk} I_{[X_{nk} \notin K]}|| < \frac{\varepsilon\delta}{8}. \qquad (6.1.10)$$

Since $Q_t(x) \to 0$ as $t \to \infty$ for each $x \in E$ and K is compact, pick t so that

$$\sup_{x \in K} \|Q_t(x)\| < \frac{\varepsilon\delta}{8}. \qquad (6.1.11)$$

For each n

$$P[\|\sum_{k=1}^{n} a_{nk}X_{nk}\| > \varepsilon]$$

$$\leq P[\|\sum_{k=1}^{n} a_{nk}U_t(X_{nk})\| > \frac{\varepsilon}{2}] + P[\|\sum_{k=1}^{n} a_{nk}Q_t(X_{nk})\| > \frac{\varepsilon}{2}]$$

$$\leq \sum_{i=1}^{t} P[|\sum_{k=1}^{n} a_{nk}f_i(X_{nk})| > \frac{\varepsilon}{2t\|b_i\|}] + \frac{2}{\varepsilon}\sum_{k=1}^{n} |a_{nk}|E\|Q_t(X_{nk})\|. \qquad (6.1.12)$$

From (6.1.10) and (6.1.11)

$$\frac{2}{\varepsilon}\sum_{k=1}^{n} |a_{nk}|E\|Q_t(X_{nk})\| \leq \frac{2}{\varepsilon}\sum_{k=1}^{n} |a_{nk}|E[\frac{\varepsilon\delta}{8} + \|X_{nk}I_{[X_n \notin K]}\|]$$

$$\leq \frac{2}{\varepsilon}[\frac{\varepsilon\delta}{8} + \frac{\varepsilon\delta}{8}] = \frac{\delta}{2} \qquad (6.1.13)$$

for each n. For each i

$$P[|\sum_{k=1}^{n} a_{nk}f_i(X_{nk})| > \frac{\varepsilon}{2t\|b_i\|}]$$

$$\leq (\frac{2t\|b_i\|}{\varepsilon})^2 [\sum_{k=1}^{n} a_{nk}^2 E[f_i^2(X_{nk})] + \rho_n(f_i)\sum_{k\neq \ell}\sum |a_{nk}a_{n\ell}|]$$

which goes to 0 as $n \to \infty$ since $\sum_{k\neq\ell}\sum |a_{nk}|\,|a_{n\ell}| \leq 1$. Thus, for all $n \geq N(\varepsilon,\delta)$

$$P[||\sum_{k=1}^{n} a_{nk}U_t(X_{nk})|| > \varepsilon/2] < \frac{\delta}{2}, \qquad (6.1.14)$$

and the result follows by (6.1.12), (6.1.13), and (6.1.14)./ / /

Exchangeability was not needed in Theorem 6.1.2. However, by assuming that for each $f \in E^*$ $\{f(X_{nk}): k \geq 1\}$ are exchangeable for each n, the condition that $\rho_n(f) \to 0$ for each $f \in E^*$ could be replaced by a condition similar to (6.1.4) on the truncated conditional means of $\{f(X_{nk})\}$. This replacement follows the arguments in Section 1.3 and Problem 1.8. In addition, $\rho_n(f) \to 0$ can imply the conditional means must be going to zero (rf: Theorem 4.2.3). Also, it is clear that the hypothesis needs only apply to the coordinate functional when E has a Schauder basis.

6.2 STRONG CONVERGENCE RESULTS FOR SEQUENCES AND INFINITE ARRAYS OF EXCHANGEABLE RANDOM ELEMENTS

In this section, strong convergence results will be obtained for sequences of exchangeable random elements in Banach spaces. Strong convergence results are also obtained for infinite arrays of random elements that are row-wise exchangeable. The first result by Bru, Heinich and Lootgieter (1981) is an application of type (s) function Recall that a function f on R, such that $f(0) = 0$, is said to be of type (s) if $f(x)/x$ is decreasing.

Definition 6.2.1 Let E and F be Banach lattices of real-valued functions. Let $f: F_+ \to E_+$. Then f is of type (s) if for all $u \in E_+^*$, $u \circ f$ is of type (s).

Recall from Section 3.2 that ϕ is a positive subadditive function.

<u>Theorem 6.2.1</u> If f is a function of type (s) from $F_+ \to E_+$ where E is weakly sequentially complete, then for each sequence of positive exchangeable random elements such that $E(||\phi(X_1)||) < +\infty$, $(1/n)\ \phi(X_1+\ldots+X_n)$ is a sub-martingale that converges almost surely in $L^1(E)$. Here, the σ-fields are simply $\sigma\{X_1, \ldots, X_n\}$.

<u>Definition 6.2.2</u> Let $\{X_{nk}: n \geq 1, k \geq 1\}$ be an array of random elements in E. The array $\{X_{nk}\}$ is said to be <u>T-exchangeable</u> (row-column exchangeable) if both $\{X_{nk}: k \in N\}$ is exchangeable for each n and if $\{X_{nk}: n \in N\}$ is exchangeable for each k.

The next result by Edgar and Sucheston (1981) is for T-exchangeable random elements in $L_p(E)$; $0 < p \leq 1$ (rf: Definition 4.1.1).

<u>Theorem 6.2.2</u> If $\{X_{nk}: n \geq 1, k \geq 1\}$ is T-exchangeable, $0 < p \leq 1$, and $||X_{1,1}||^p \in L \log L$, then $(mn)^{-1/p} \sum_{\substack{i \leq m \\ j \leq n}} X_{ij}$ converges almost surely in $L_p(E)$. If $p < 1$, the limit X is zero. If $p = 1$, $X = E[X_{11}|\mathcal{E}]$ where $\mathcal{E}$ is the σ-field of T-exchangeable events.

The next three results by Taylor (to appear) are three strong laws of large numbers for infinite arrays of exchangeable random elements that are row-wise exchangeable. In these three results, the rate at which the measure of non-orthogonality, ρ_n, goes to zero will play a crucial role. In particular, the methods of proof will yield the stronger complete convergence rather than the traditional almost sure convergence. The random elements $\{S_n\}$ are said to converge completely to 0 if $\sum_{n=1}^{\infty} P[||S_n|| > \varepsilon] < \infty$ for each $\varepsilon > 0$. The first result is a strong law for triangular arrays.

<u>Theorem 6.2.3</u> Let $\{X_{nk}\}$ be an array of random elements in a separable Banach space of type p, $1 < p \leq 2$, and let $\{X_{nk}: k \geq 1\}$ be exchangeable for each n. If for each $\eta > 0$

$$\sum_{n=1}^{\infty} \mu_n(\{P_\nu: ||E_\nu(X_{n1})|| > \eta\}) < \infty \qquad (6.2.1)$$

and for some $q \geq 1$

$$\sum_{n=1}^{\infty} E(||X_{n1}||^{pq})/n^{q(p-1)} < \infty, \qquad (6.2.2)$$

then

$$||\frac{1}{n} \sum_{k=1}^{n} X_{nk}|| \to 0 \quad \text{completely.}$$

<u>Proof</u> (in outline form): For each $\varepsilon > 0$ and each n

$$P[||\frac{1}{n} \sum_{k=1}^{n} X_{nk}|| > \varepsilon] = \int_F P_\nu[||\frac{1}{n} \sum_{k=1}^{n} X_{nk}|| > \varepsilon] d\mu_n(P_\nu)$$

$$\leq \int_F P_\nu[||\frac{1}{n} \sum_{k=1}^{n} X_{nk} - E_\nu X_{n1}|| > \frac{\varepsilon}{2}] d\mu_n(P_\nu)$$

$$+ \int_F P_\nu[||E_\nu X_{n1}|| > \frac{\varepsilon}{2}] d\mu_n(P_\nu)$$

$$\leq \int_F n^{-pq}\left(\frac{\varepsilon}{2}\right)^{-pq} C\, n^q E_\nu(||X_{n1} - E_\nu X_{n1}||^{pq})\,d\mu_n(P_\nu)$$

$$+ \mu_n(\{P_\nu : ||E_\nu X_{n1}|| > \frac{\varepsilon}{2}\})$$

by Corollary 2.1 of Woycyznski (1980) where C is related to the type p condition and does not depend on n. Since

$$\int_F E_\nu(||X_{n1} - E_\nu X_{n1}||^{pq})\,d\mu_n(P_\nu)$$

$$\leq 2^{pq} \int_F E_\nu(||X_{n1}||^{pq})\,d\mu_n(P_\nu) = 2^{pq} E(||X_{n1}||^{pq}),$$

it follows that

$$\sum_{n=1}^{\infty} P[||\frac{1}{n}\sum_{k=1}^{n} X_{nk}|| > \varepsilon] \leq \left(\frac{4}{\varepsilon}\right)^{pq} C \sum_{n=1}^{\infty} E(||X_{n1}||^{pq})/n^{q(p-1)}$$

$$+ \sum_{n=1}^{\infty} \mu_n(\{P_\nu : ||E_\nu X_{n1}|| > \frac{\varepsilon}{2}\}) < \infty$$

from (6.2.1) and (6.2.2). ///

Similar to (1.3) of Taylor (to appear), it can be shown that $\sum_{n=1}^{\infty} \rho_n < \infty$ implies that Condition (6.2.1) holds. However, it is easy to see that the converse is not true by letting $X_{nk} \equiv 1/\sqrt{n}$. The type p, $p > 1$, condition is also necessary in Theorem 6.2.3 since Beck (1963) showed that B-convexity ($\equiv$ type p for some $p > 1$) is necessary and sufficient for the strong law of large numbers to hold for sequences of independent, identically distributed, uniformly bounded random elements with zero means. By assuming that the random elements are uniformly bounded, a strong law can be obtained for weighted sums of random elements with more general weights. The proof of Theorem 6.2.4 is similar to the proof of Theorem 6.2.3 and will not be presented.

Theorem 6.2.4 Let $\{X_{nk}\}$ be an array of random elements in a separable B-convex space such that for some constant K $||X_{nk}|| \leq K$ a.s. for each n and k such that $\{X_{nk}: k \geq 1\}$ are exchangeable for each n. Let $\{a_{nk}\}$ be constants such that

$$\sum_{k=1}^{\infty} |a_{nk}| \leq 1 \text{ and } \max_k |a_{nk}| = o(n^{-\gamma})$$

for some $\gamma > 0$. If Condition (6.2.1) holds, then $||\sum_{k=1}^{\infty} a_{nk}X_{nk}|| \to 0$ completely.

For spaces which are not B-convex, that is, type 1 spaces, a tightness condition is generally needed (as previously indicated). The last result requires that the random elements be restricted to a compact set which is equivalent to a uniform boundedness condition in finite-dimensional spaces. Thus, the generality of the weights in Theorem 6.2.5 provides for a very applicable strong law. Theorem 6.2.5 can be proven in a similar manner to that of the proof of Theorem 6.2.2 and will thus be presented without proof.

Theorem 6.2.5 Let $\{X_{nk}\}$ be random elements in a separable Banach space E such that for some compact set K $P[X_{nk} \in K] = 1$ for all n and k. Let $\{X_{nk}: k \geq 1\}$ be exchangeable for each n and let $\{a_{nk}\}$ be constants such that $\sum_{k=1}^{\infty} |a_{nk}| \leq 1$ and $\sum_{n=1}^{\infty} \exp(-\alpha/A_n) < \infty$ for each $\alpha > 0$ where $A_n = \sum_{k=1}^{\infty} a_{nk}^2$. If Condition (6.2.1) holds, then

$||\sum_{k=1}^{\infty} a_{nk}X_{nk}|| \to 0$ completely.

Remark: If $\max_k |a_{nk}| \le Bn^{-\gamma}$ for some $\gamma > 0$, then $A_n \le Bn^{-\gamma}$ and $\sum_{n=1}^{\infty} \exp(-\alpha/A_n) \le \sum_{n=1}^{\infty} \exp(-\alpha B^{-1}n^{\gamma}) < \infty$.

6.3 STRONG CONVERGENCE RESULTS FOR TRIANGULAR ARRAYS OF EXCHANGEABLE RANDOM ELEMENTS

In this section, three strong laws of large numbers for triangular arrays of random elements which are row-wise exchangeable will be presented. Using Lemmas 4.3.2 and 4.3.3, a strong law of large numbers will be developed for triangular arrays of random elements which are tight and row-wise exchangeable. It is well-established that for separable (infinite-dimensional) Banach spaces a tightness condition (in lieu of geometric conditions on the space) is required (rf: Chapter 4 of Taylor (1978) and specifically Example 4.1.1). In Chapter 4 a general condition involving tightness of distributions and moments of random elements $\{X_{nk}\}$ was stated as:

Given $\varepsilon > 0$, there exists a compact subset K such that

$$\sup_n (E||X_{n1}I_{[X_{n1} \notin K]}||) < \varepsilon \tag{6.3.1}$$

Theorem 6.3.1 Let $\{X_{nk}: 1 \le k \le n, n \ge 1\}$ be an array of random elements in a separable Banach space E. Let $\{X_{nk}\}$ be row-wise exchangeable such that (6.3.1) holds. Let F_n be the σ-field in (4.3.1). If for each $f \in E^*$

(i) $\{E(f(X_{n1})|F_n): n \ge 1\}$ is a reverse martingale,

(ii) $E[f(X_{n1})f(X_{n2})] \to 0$ as $n \to \infty$,

(iii) $E(f^2(X_{n1})) = o(n)$ for $f \in E^*$,

(iv) $\{E(X_{n1} I_{[X_{n1} \notin K]} | \mathcal{F}_n): n \geq 1\}$ is a reverse martingale for each compact subset K,

then

$$\left|\left|\frac{1}{n} \sum_{k=1}^{n} X_{nk}\right|\right| \to 0 \text{ a.s.}$$

Proof: Let $\varepsilon > 0$, and $\eta > 0$ be given. Let K, without loss of generality, be a convex, symmetric compact set such that (6.3.1) holds. Since K is compact, there exist continuous linear functionals $f_1, f_2, \ldots, f_t \in E^*$ such that $||f_i|| = 1$ for $1 \leq i \leq t$ and

$$\{x \in K: ||x|| > \eta/2\} \subset \bigcup_{i=1}^{t} \{x \in K: |f_i(x)| > \eta/4\}$$
$$\subset \bigcup_{i=1}^{t} \{x: |f_i(x)| > | \eta/4\} \qquad (6.3.2)$$

Since $\frac{1}{n} \sum_{k=1}^{n} X_{nk} I_{[X_{nk} \in K]} \in K$ for each n,

$$P[\sup_{n \geq m} ||\frac{1}{n} \sum_{k=1}^{n} X_{nk}|| > \eta].$$
$$\leq P[\sup_{n \geq m} ||\frac{1}{n} \sum_{k=1}^{n} X_{nk} I_{[X_{nk} \in K]}|| > \eta/2]$$
$$+ P[\sup_{n \geq m} ||\frac{1}{n} \sum_{k=1}^{n} X_{nk} I_{[X_{nk} \notin K]}|| > \eta/2]$$
$$\leq P[\bigcup_{i=1}^{t} [\sup_{n \geq m} |\frac{1}{n} \sum_{k=1}^{n} f_i(X_{nk})| > \eta/4]]$$

$$+ P[\sup_{n \geq m} ||\frac{1}{n} \sum_{k=1}^{n} X_{nk} I_{[X_{nk} \notin K]}|| > \eta/2]$$

$$\leq \sum_{i=1}^{t} P[\sup_{n \geq m} |\frac{1}{n} \sum_{k=1}^{n} f_i(X_{nk})| > \eta/4]$$

$$+ P[\sup_{n \geq m} ||\frac{1}{n} \sum_{k=1}^{n} X_{nk} I_{[X_{nk} \notin K]}|| > \eta/2]. \quad (6.3.3)$$

Note that $\{f(X_{nk}): 1 \leq k \leq n\}$ and $\{X_{nk} I_{[X_{nk} \notin K]}: 1 \leq k \leq n\}$ are exchangeable for each n. Using Theorem 3.3.1,

$$P[\sup_{n \geq m} |\frac{1}{n} \sum_{k=1}^{n} f_i(X_{nk})| > \eta/4] \to 0 \text{ as } m \to \infty.$$

Also, by (4.3.1) and (4.3.2)

$$P[\sup_{n \geq m} ||\frac{1}{n} \sum_{k=1}^{n} X_{nk} I_{[X_{nk} \notin K]}|| > \eta/2]$$

$$= P[\sup_{n \geq m} ||E(X_{n1} I_{[X_{n1} \notin K]} | \mathcal{F}_n)|| > \eta/2]$$

$$\leq \frac{2}{\eta} \int_{[\sup_{n \geq m} ||E(X_{n1} I_{[X_{n1} \notin K]} | \mathcal{F}_n)|| > \eta/2]} ||E(X_{m1} I_{[X_{m1} \notin K]} | \mathcal{F}_m)|| dP$$

$$\leq \frac{2}{\eta} \int_{\Omega} ||E(X_{m1} I_{[X_{m1} \notin K]} | \mathcal{F}_m)|| dP.$$

For each m, (6.3.1) provides that

$$\frac{2}{\eta} \int_{\Omega} ||E(X_{m1} I_{[X_{m1} \notin K]} | \mathcal{F}_m)|| dP$$

$$\leq \frac{2}{\eta} \int_{\Omega} E(||X_{m1} I_{[X_{m1} \notin K]}|| \,|F_m)dP$$

$$= \frac{2}{\eta} E||X_{m1} I_{[X_{m1} \notin K]}||$$

$$\leq \frac{2}{\eta} \varepsilon. \tag{6.3.4}$$

Thus, from (6.3.2) and (6.3.4)

$$P[\sup_{n \geq m} ||\frac{1}{n} \sum_{k=1}^{n} X_{nk}|| > \eta]$$

$$\leq \sum_{i=1}^{t} P[\sup_{n \geq m} |\frac{1}{n} \sum_{k=1}^{n} f_i(X_{nk})| > \eta/4] + 2\varepsilon/\eta$$

$\to 0 + 2\varepsilon/\eta$ as $m \to \infty$ since t is finite and fixed. Also since ε is arbitrary, it follows that

$P[\sup_{n \geq m} ||\frac{1}{n} \sum_{k=1}^{n} X_{nk}|| > \eta] \to 0$ as $m \to \infty$. Hence

$||\frac{1}{n} \sum_{k=1}^{n} X_{nk}|| \to 0$ a.s. ///

If the Banach space has a Schauder basis $\{b_i\}$ and coordinate functionals $\{f_i\}$, then Conditions (i), (ii), and (iii) need only hold for each coordinate functional f_i. This can be seen by letting $U_t(x) = \sum_{i=1}^{t} f_i(x)b_i$ and $Q_t(x) = x - U_t(x)$ be the continuous, linear partial sum operator and remainder term, respectively. From Lemma 1.3.3 of Taylor (1978) $\sup_{x \varepsilon k} ||x - U_t(x)|| \to 0$ as $t \to \infty$ for each compact set K which establishes (6.3.2) in the proof of Theorem 6.3.1. This technique also allows a relaxing of Condition (iv). Specifically, every separable Banach space can be embedded isometrically in a Banach space which has a Schauder basis. Moreover, it can be assumed that the basis is monotone so that $||U_t(x)|| \leq ||x||$ and $||Q_t(x)|| \leq ||x||$ for each $x \in E$ and $t \geq 1$. Theorem 6.3.2 summarizes these results and is the desired version of Theorem 3.3.1 for Banach spaces.

Theorem 6.3.2 Let $\{X_{nk}: 1 \leq k \leq n, n \geq 1\}$ be an array of random elements in a separable Banach space E such that Condition (6.3.1) holds. Let $\{X_{nk}: 1 \leq k \leq n\}$ be exchangeable for each n. If

(i) $\{E(X_{n1}|\mathcal{F}_n): n \geq 1\}$ is a reverse martingale,

(ii) $E[f(X_{n1})f(X_{n2})] \to 0$ for each $f \in E^*$

(iii) $E[f(X_{n1})^2] = o(n)$ for each $f \in E^*$,

then

$$||\frac{1}{n}\sum_{k=1}^{n} X_{nk}|| \to 0 \text{ a.s.}$$

Proof: The isometric embedding of E into a Banach space which has a monotone (Schauder) basis preserves exchangeability, (6.3.1) and Conditions (i), (ii), and (iii). Thus, assuming that E* has a monotone basis, let $\varepsilon > 0$ and $\eta > 0$ be given and let the compact subset K of (6.3.1) be convex and symmetric. Choose t so that

$$\sup_{x \in K} ||Q_t(x)|| < \varepsilon. \tag{6.3.5}$$

Next, for each m

$$P[\sup_{n \geq m} ||\frac{1}{n}\sum_{k=1}^{n} X_{nk}|| > \eta]$$

$$\leq P[\sup_{n\geq m} ||\frac{1}{n}\sum_{k=1}^{n} U_t(X_{nk})|| > \eta/2]$$

$$+ P[\sup_{n\geq m} ||\frac{1}{n}\sum_{k=1}^{n} Q_t(X_{nk})|| > \eta/2]. \qquad (6.3.6)$$

First,

$$P[\sup_{n\geq m} ||\frac{1}{n}\sum_{k=1}^{n} U_t(X_{nk})|| > \eta/2]$$

$$\leq \sum_{i=1}^{t} P[\sup_{n\geq m} |\frac{1}{n}\sum_{k=1}^{n} f_i(X_{nk})| > \eta/2t||b_i||]$$

which goes to 0 as $m \to \infty$ following the same steps as the proof of Theorem 6.3.1. From Lemma 4.3.3 $\{E(Q(X_{n1})|F_n): n \geq 1\}$ is a reverse martingale. Since $\{Q_t(X_{nk}): 1 \leq k \leq n\}$ are exchangeable random elements, Property 1.2.4 and (4.3.2) provides

$$P[\sup_{n\geq m} ||\frac{1}{n}\sum_{k=1}^{n} Q_t(X_{nk})|| > \eta/2]$$

$$= P[\sup_{n\geq m} ||E(Q_t(X_{n1})|F_n)|| > \eta/2]$$

$$\leq \frac{2}{\eta} \int_{[\sup_{n\geq m} ||E(Q_t(X_{n1})|F_n)|| > \eta/2]} ||E(Q_t(X_{m1})|F_m)||dP$$

$$\leq \frac{2}{\eta} E||E(Q_t(X_{m1})|F_m)|| \leq \frac{2}{\eta} E||Q_t(X_{m1})||. \qquad (6.3.7)$$

But from (6.3.5) and the monotone basis property

$$E||Q_t(X_{m1})|| \leq E||Q_t(X_{m1}I_{[X_{m1} \varepsilon K]})||+E||Q_t(X_{m1}I_{[X_{m1} \notin K]})||$$

$$\leq E(\varepsilon) + E||X_{m1}I_{[X_{m1} \notin K]}|| < 2\varepsilon \qquad (6.3.8)$$

for each m. Combining (6.3.6), (6.3.7) and (6.3.8), it follows that

$$\limsup_{m\to\infty} P[\sup_{n\geq m} ||\frac{1}{n} \sum_{k=1}^{n} X_{nk}|| > \eta] \leq 0 + 4\varepsilon/\eta,$$

and the proof is complete since ε is arbitrary. ///

Several previous results required that $\sum_{n=1}^{\infty} E(X_{n1}X_{n2}) < \infty$, infinite exchangeability and more restrictive conditions of the moments $\{X_{n1}: n \geq 1\}$. But, they did not use martingale techniques and did not require that $\{E(X_{n1}|F_n): n \geq 1\}$ be a reverse martingale.

The strong law of large numbers is for arbitrary Banach spaces. A preliminary result is needed to obtain the next strong law of large numbers. This result will be presented without proof.

<u>Lemma 6.3.3</u> Let $\{X_{nk}\}$ be an array of random elements in a separable Banach space. Let

$$U_{nn} = \sigma\{\sum_{k=1}^{n} X_{nk}, \sum_{k=1}^{n+1} X_{(n+1),k}, \ldots\} \qquad (6.3.9)$$

and

$$U_{\infty n} = \sigma\{\sigma\{\sum_{k=1}^{n} X_{\infty k}, \sum_{k=1}^{n+1} X_{\infty k}, \ldots\} \cup U_{nn}\} \qquad (6.3.10)$$

such that $||X_{nk} - X_{\infty k}|| \to 0$ in mean for each k. If $||X_{nk} - X_{\infty k}|| \geq ||X_{(n+1),k} - X_{\infty k}||$ for all n, then $E(||X_{nk} - X_{\infty k}|| \mid U_{\infty n}) \to 0$ a.s.

<u>Theorem 6.3.4</u> Let $\{X_{nk}: 1 \leq k \leq n, n \geq 1\}$ be an array of random elements in a separable Banach space E. Let $\{X_{nk}\}$ be row-wise exchangeable for each n and let U_{nn} and $U_{\infty n}$ be the σ-fields defined in (6.3.9) and (6.3.10). Let $\{X_{nk}\}$ converge in the second mean for each k and let $||X_{n1} - X_{\infty 1}|| \geq ||X_{(n+1),1} - X_{\infty 1}||$ a.s. for each n. If $\rho_n(f) = E[f(X_{n1})f(X_{n2})] \to 0$ as $n \to \infty$ for each $f \in E^*$, then $||\frac{1}{n}\sum_{k=1}^{n} X_{nk}|| \to 0$ a.s.

<u>Proof</u>: Note the following

(i) $U_{nn} \supseteq U_{(n+1),(n+1)}$; $U_{\infty n} \supseteq U_{nn}$.

(ii) $E(X_{n1}|U_{nn}) = E(X_{n1}|U_{\infty n}) = \frac{1}{n}\sum_{k=1}^{n} X_{nk}$;

$$E(X_{\infty 1}|U_{\infty n}) = \frac{1}{n}\sum_{k=1}^{n} X_{\infty k}.$$

(iii) $\{E(X_{\infty 1}|U_{\infty n})\ n \geq 1\}$ is a reverse martingale.

By Theorem 4.2.2 $\{X_{\infty k}: k \geq 1\}$ is an exchangeable sequence of random elements. Thus, for $\varepsilon > 0$, and by (4.3.1)

$$P[\sup_{n \geq m} ||\frac{1}{n}\sum_{k=1}^{n} X_{nk}|| > \varepsilon]$$

$$\leq P[\sup_{n\geq m} ||\frac{1}{n}\sum_{k=1}^{n} X_{nk} - \frac{1}{n}\sum_{k=1}^{n} X_{\infty k}|| > \frac{\varepsilon}{2}]$$

$$+ P[\sup_{n\geq m} ||\frac{1}{n}\sum_{k=1}^{n} X_{\infty k}|| > \frac{\varepsilon}{2}]$$

$$= P[\sup_{n\geq m} ||E(X_{n1}|U_{\infty n}) - E(X_{\infty 1}|U_{\infty n})|| > \frac{\varepsilon}{2}]$$

$$+ P[\sup_{n\geq m} ||\frac{1}{n}\sum_{k=1}^{n} X_{\infty k}|| > \frac{\varepsilon}{2}]$$

$$= P[\sup_{n\geq m} ||E(X_{n1} - X_{\infty 1}|\mathcal{F}_{\infty n})|| > \frac{\varepsilon}{2}]$$

$$+ P[\sup_{n\geq m} ||\frac{1}{n}\sum_{k=1}^{n} X_{\infty k}|| > \frac{\varepsilon}{2}].$$

Now by (6.1.1)

$$P[\sup_{n\geq m} ||\frac{1}{n}\sum_{k=1}^{n} X_{\infty k}|| > \frac{\varepsilon}{2}]$$

$$= \int_F P_\nu[\sup_{n\geq m} ||\frac{1}{n}\sum_{k=1}^{n} X_{\infty k}|| > \frac{\varepsilon}{2}]d\mu_\infty(P_\nu)$$

where $\{X_{\infty k}: k \geq 1\}$ are independent, identically distributed random elements with respect to P_ν. By Theorem 4.2.3 $E_\nu(X_{\infty 1}) = 0$, and it follows from Mourier's strong law of large numbers for independent, identically distributed random elements that $||\frac{1}{n}\sum_{k=1}^{n} X_{\infty k}|| \to 0$ a.s. as $n \to \infty$.

Thus,

$$P_\nu[\sup_{n\geq m} ||\frac{1}{n}\sum_{k=1}^{n} X_{\infty k}|| > \frac{\varepsilon}{2}] \to 0 \text{ as } n \to \infty.$$

Hence, by the Lebesgue dominated convergence theorem,

$$\int_F P_\nu[\sup_{n\geq m} ||\frac{1}{n}\sum_{k=1}^{n} X_{\infty k}|| > \frac{\varepsilon}{2}]d\mu_\infty(P_\nu) \to 0 \text{ as } n \to \infty .$$

By Lemma 6.3.3 ,

$$E(||X_{n1} - X_{\infty 1}|| \; |F_{\infty n}) \overset{a.s.}{\to} 0 \text{ as } n \to \infty.$$

Thus,

$$P[\sup_{n\geq m} ||E(X_{n1} - X_{\infty 1}|F_{\infty n})|| > \frac{\varepsilon}{2}]$$

$$\leq P[\sup_{n\geq m} E(||X_{n1} - X_{\infty 1}|| \; |F_{\infty n}) > \frac{\varepsilon}{2}] \to 0 \text{ as } n \to \infty.$$

Thus, it follows that

$$P[\sup_{n\geq m} ||\frac{1}{n}\sum_{k=1}^{n} X_{nk}|| < \varepsilon] \to 0 \text{ as } n \to \infty.$$

Hence,

$$||\frac{1}{n}\sum_{k=1}^{n} X_{nk}|| \to 0 \text{ a.s. } \text{ as } n \to \infty. \qquad ///$$

The measure of non-orthogonality condition in Lemma 6.3.3 is, in a sense, quite weaker than the condition $\sum_{n=1}^{\infty} \mu_n(\{P_\nu: ||E_\nu(X_{n1})|| > \eta\}) < \infty$ used in obtaining a strong law of large numbers by Taylor (to appear). That is, the condition $\sum_{n=1}^{\infty} \rho_n < \infty$ relates to $\sum \mu_n(\{P_\nu: ||E_\nu(X_{n1})|| > \eta\}) < \infty$ whereas the weak laws used a condition similar to $\rho_n \to 0$ as $n \to \infty$.

The next two theorems and corollary by Woyczynski (1980) deal with independent, identically distributed random elements with zero means in type p Banach spaces. These will be listed and used to obtain a strong law of large numbers for row-wise exchangeable random elements in type p Banach spaces (rf: Theorem 4.1.2).

Theorem 6.3.5 Let $1 \leq p \leq 2$ and $q \geq 1$. The following properties of E are equivalent:

(i) E is of type p

(ii) There exists a C such that for every $n \in N$ and for any sequence $\{X_i\}$ of independent random elements in E with $EX_i = 0$

$$E\|\sum_{i=1}^{n} X_i\|^q \leq CE(\sum_{i=1}^{n} \|X_i\|^p)^{q/p}. \qquad (6.3.11)$$

If $q = p$, (6.3.11) becomes $E\|\sum_{k=1}^{n} X_i\|^p \leq CnE\|X_1\|^p$.

Corollary 6.3.6 Let E be of type p and $q \geq p$. If $\{X_n\}$ are independent, identically distributed random elements in E with $E(X_1) = 0$, then

$$E\|\sum_{i=1}^{n} X_i\|^q \leq Cn^{q/p}E\|X_1\|^q. \qquad (6.3.12)$$

The next result by Woyczynski (1980) is for a sequence of zero mean, independent random elements in E with uniformly bounded tail probabilities.

Theorem 6.3.7 Let $1 < p < 2$. The following properties of a Banach space E are equivalent:

(i) E is type $p + \delta$

(ii) For any sequence $\{X_i\}$ of zero mean, independent random elements in E with uniformly bounded tail probabilities, the series $\sum_{n=1}^{\infty} X_n/n^{1/p}$ converges a.s.

(iii) For any sequence $\{X_i\}$ as in (ii) $\|\sum_{k=1}^{n} X_k/n^{1/p}\| \to 0$ a.s.

Since $\{X_{\infty k}: k \geq 1\}$ are exchangeable, they are identically distributed. Hence, they have uniformly bounded tail probabilities. That is, for all $t \in R$ and $k \geq 1$

$$P_\nu(\|X_{\infty k}\| > t) \leq P_\nu(\|X_{\infty 1}\| > t) \qquad (6.3.13)$$

and a strong law of large numbers for row-wise exchangeable random elements in type p separable Banach spaces can be obtained.

<u>Theorem 6.3.8</u> Let $\{X_{nk}: 1 \leq k \leq n, n \geq 1\}$ be an array of random elements in a separable Banach space of type p, $1 \leq p < 2$. Let $\{X_{nk}\}$ be exchangeable for each n and let U_{nn} and $U_{\infty n}$ be the σ-fields defined in (6.3.9) and (6.3.10). If

(i) $\|X_{n1} - X_{\infty 1}\| \geq \|X_{(n+1)1} - X_{\infty 1}\|$ for each n

(ii) $E\|X_{nk} - X_{\infty k}\|^2 = o(n^{-2\alpha})$, where $\alpha = (p-1)/p$, and

(iii) $\rho_n(f) = E[f(X_{n1})f(X_{n2})] \to 0$ as $n \to \infty$ for $f \in E^*$,

then

$$\left\|\frac{1}{n^{1/p}} \sum_{k=1}^{n} X_{nk}\right\| \to 0 \text{ a.s. as } n \to \infty.$$

Proof: Since $E\|X_{n1} - X_{\infty 1}\|^2 = o(n^{-2\alpha})$, then $X_{nk} \to X_{\infty k}$ in the 2nd mean. By (4.3.1)

(i) $E(X_{n1}|U_m) = E(X_{n1}|U_{\infty n}) = \frac{1}{n} \sum_{k=1}^{n} X_{nk}$ a.s., and

(ii) $E(X_{\infty 1}|U_{\infty n}) = \frac{1}{n} \sum_{k=1}^{n} X_{\infty k}$ a.s.

and since $U_{\infty n} \supseteq U_{\infty n+1}$, $\{E(X_{\infty 1}|U_{\infty n})\}$ is a reverse martingale. By Theorem 4.2.2, $\{X_{\infty k}: k \geq 1\}$ is an exchangeable sequence of random elements. Thus for $\varepsilon > 0$, and by (4.3.1)

$$P[\sup_{n\geq m} \left\|\frac{1}{n^{1/p}} \sum_{k=1}^{n} X_{nk}\right\| > \varepsilon]$$

$$\leq P[\sup_{n\geq m} \left\|\frac{1}{n^{1/p}} (\sum_{k=1}^{n} X_{nk} - \sum_{k=1}^{n} X_{\infty k})\right\| > \frac{\varepsilon}{2}]$$

$$+ P[\sup_{n\geq m} \left\|\frac{1}{n^{1/p}} \sum_{k=1}^{n} X_{\infty k}\right\| > \frac{\varepsilon}{2}]$$

$$= P[\sup_{n\geq m} \left\|n^{\frac{p-1}{p}} (\frac{1}{n} \sum_{k=1}^{n} X_{nk} - \frac{1}{n} \sum_{k=1}^{n} X_{\infty k})\right\| > \frac{\varepsilon}{2}]$$

$$+ P[\sup_{n>m} \left\|\frac{1}{n^{1/p}} \sum_{k=1}^{n} X_{\infty k}\right\| > \frac{\varepsilon}{2}]$$

$$= P[\sup_{n\geq m} \left\|n^{\frac{p-1}{p}} E(X_{n1} - X_{\infty 1}|F_{\infty n})\right\| > \frac{\varepsilon}{2}]$$

$$+ P[\sup_{n\geq m} \left\|\frac{1}{n^{1/p}} \sum_{k=1}^{n} X_{\infty k}\right\| > \frac{\varepsilon}{2}].$$

By (6.1.1)

$$P[\sup_{n\geq m} ||\frac{1}{n^{1/p}} \sum_{k=1}^{n} X_{\infty k}|| > \frac{\varepsilon}{2}]$$

$$= \int_{F} P_{\nu}[\sup_{n\geq m} ||\frac{1}{n^{1/p}} \sum_{k=1}^{n} X_{\infty k}|| > \frac{\varepsilon}{2}] d\mu_{\infty}(P_{\nu}),$$

where $\{X_{\infty k}: k \geq 1\}$ are independent, and identically distributed with respect to P_{ν}. By Theorem 4.2.3, $E_{\nu}X_{\infty 1} = 0$. Thus, by Theorem 6.3.7

$$||\frac{1}{n^{1/p}} \sum_{k=1}^{n} X_{\infty k}|| \to 0 \text{ a.s.}$$

which implies that $P_{\nu}[\sup_{n\geq m} ||\frac{1}{n^{1/p}} \sum_{k=1}^{n} X_{\infty k}|| > \frac{\varepsilon}{2}] \to 0$ as $n \to \infty$.

By the Lebesgue dominated convergence theorem,

$$\int_{F} P_{\nu}[\sup_{n\geq m} ||\frac{1}{n^{1/p}} \sum_{k=1}^{n} X_{\infty k}|| > \frac{\varepsilon}{2}] d\mu_{\infty}(P_{\nu}) \to 0 \text{ as } n \to \infty.$$

For $B \in F_{\infty n}$ and $\alpha = \frac{p-1}{p}$, by Hölder's inequality

$$\int_{B} n^{\alpha} ||E(X_{n1} - X_{\infty 1}|F_{\infty n})|| dP \leq \int_{B} n^{\alpha} E(||X_{n1} - X_{\infty 1}|| \, |F_{\infty n}) dP$$

$$\leq \int_{\Omega} n^{\alpha} E(||X_{n1} - X_{\infty 1}|| \, |F_{\infty n}) dP$$

$$= n^{\alpha} E(||X_{n1} - X_{\infty 1}||)$$

$$\leq [n^{2\alpha} E(||X_{n1} - X_{\infty 1}||^{2})]^{1/2}$$

$$\to 0 \text{ as } n \to \infty$$

by (ii). Hence, by Lemma 6.3.2

$$n^{\alpha}||E(X_{n1} - X_{\infty 1}|F_{\infty n})|| \to 0 \text{ a.s.}$$

This implies that

$$P[\sup_{n \geq m} ||n^{\frac{p-1}{p}} E(X_{n1} - X_{\infty 1}|U_{nn})|| > \frac{\varepsilon}{2}] \to 0 \text{ as } n \to \infty.$$

Hence,

$$P[\sup_{n \geq m} ||n^{-1/p} \sum_{k=1}^{n} X_{nk}|| > \varepsilon] \to 0 \text{ as } n \to \infty.$$

Thus,

$$||n^{-1/p} \sum_{k=1}^{n} X_{nk}|| \to 0 \text{ a.s.} \qquad ///$$

Remark: It should be noted that if $||X_{n1} - X_{\infty 1}|| \to$ a.s., then the condition $||X_{n1} - X_{\infty 1}|| \geq ||X_{(n+1),1} - X_{\infty 1}||$ for all n is not needed in the proof of Theorem 6.3.4. That is, $||X_{n1} - X_{\infty 1}|| \to 0$ a.s. implies that $||E(X_{n1} - X_{\infty 1}|U_{\infty n})|| \to 0$ a.s. which is crucial to the proof of Theorem 6.3.4. It is sometimes easier to show directly that $||E(X_{n1} - X_{\infty 1}|U_{\infty n})|| \to 0$ a.s. as will be demonstrated in the kernel density estimation example to follow.

The following example deals with the general density estimation problem where $X_1, \ldots, X_n$ are independent, identically distributed random variables with common probability density function f.

Example 6.3.1 Let $\{X_1, \ldots, X_n\}$ be independent identically distributed random variables with common density function f. The kernel estimator for f with constant bandwidths h_n is given by

$$f_n(t) = \frac{1}{nh_n} \sum_{k=1}^{n} K(\frac{t - X_k}{h_n})$$

where K is a bounded (integrable) kernel with compact support [a,b] and h_n is bounded and monotonically decreasing such that $h_n \to 0$ as $n \to \infty$. For additional background material, see DeVroye and Wagner (1979) and Taylor (1982b).

Let $E = \{g|g: R \to R \text{ and } ||g|| = (\int_{-\infty}^{\infty} |g(t)|^p dt)^{1/p} < \infty\}$.

Thus, E is a separable Banach space of type min{2,p}. For real numbers x and h, define K: R → E by $K(\frac{t-x}{n})$ where K is bounded and piecewise continuous. If $x_m \to x_0$, then $K(\frac{t-x_m}{h}) \to K(\frac{t-x_0}{h})$ on each piece. Thus,

$$\lim_{m\to\infty} ||K(\frac{t-x_m}{h}) - K(\frac{t-x_0}{h})||^p$$

$$= \lim_{m\to\infty} \int |K(\frac{t-x_m}{h}) - K(\frac{t-x_0}{h})|^p \, dt$$

$$= \int_{-\infty}^{\infty} \lim_{m\to\infty} |K(\frac{t-x_m}{h}) - K(\frac{t-x_0}{h})|^p \, dt = 0$$

which implies that K is continuous and hence Borel measurable. Next, define K: $R^n \to E^n$ by $(\frac{x_1}{h_n}, \ldots, \frac{x_n}{h_n}) \to$ $(K(\frac{t-x_1}{h_n}), \ldots, K(\frac{t-x_n}{h_n}))$ which is a continuous mapping of R^n into E^n. For constant h_n, $(\frac{X_1}{h_n}, \ldots, \frac{X_n}{h_n})$ is exchangeable if $(X_1, \ldots, X_n)$ is exchangeable. By Proposition 4.1.1

$$(K(\frac{t-X_1}{h_n}), \ldots, K(\frac{t-X_n}{h_n})) \qquad (6.3.14)$$

is exchangeable.

When $X_1, \ldots, X_n$ are independent, identically distributed random variables, it is clear that for any Borel measurable function K,

$$\{\frac{1}{h_n} K(\frac{t-X_1}{h_n}), \ldots, \frac{1}{h_n} K(\frac{t-X_n}{h_n})\} \text{ are exchangeable.}$$

The next proposition will prove directly that $||E(X_{n1}|U_{\infty n})|| \to 0$ a.s. In this setting $X_{\infty k} = 0$ a.s. for each k.

<u>Proposition 6.3.9</u> Let $X_{nk} = \frac{1}{h_n}(K(\frac{t-X_k}{h_n}) - E[K(\frac{t-X_1}{h_n})])$ then $||E(X_{n1}|U_{\infty n})|| \to 0$ completely (and hence almost surely).

<u>Proof</u>: Since $X_{\infty k} = 0$ a.s. for each k, $U_{nn} = \sigma\{\sum_{k=1}^{n} X_{nk}, \sum_{k=1}^{n+1} X_{(n+1),k}, \ldots\} = U_{\infty n}$. By (4.3.1), $E(X_{n1}|U_{nn}) = \frac{1}{n}\sum_{k=1}^{n} X_{nk}$ a.s. For $\varepsilon > 0$ and $q \geq 1$, by Markov's inequality and Tonelli's theorem,

$$P[||E(X_{n1}|U_{nn})|| > \varepsilon] \leq \varepsilon^{-2q} E||E(X_{n1}|U_{nn})||^{2q}$$

$$= \varepsilon^{-2q} E[(\int_{-\infty}^{\infty} | \frac{1}{nh_n} \sum_{k=1}^{n} (K(\frac{t-X_k}{h_n}) - E(K(\frac{t-X_1}{h_n})))|^p dt)^{1/p}]^{2q}$$

$$\leq \varepsilon^{-2q} E[\int_{-\infty}^{\infty} | \frac{1}{n} \sum_{k=1}^{n} \frac{1}{h_n} K(\frac{t-X_k}{h_n}) - E(\frac{1}{h_n} K(\frac{t-X_1}{h_n}))|^{2q} dt]$$

$$\leq \varepsilon^{-2q} \int_{-\infty}^{\infty} E| \frac{1}{n} \sum_{k=1}^{n} \frac{1}{h_n} K(\frac{t-X_k}{h_n}) - E(\frac{1}{h_n} K(\frac{t-X_1}{h_n}))|^{2q} dt]$$

(using that R is of type 2 and (2.8) of Taylor (1982b)

$$\leq \varepsilon^{-2q} \int_{-\infty}^{\infty} n^q n^{-2q} E\left(\left|\frac{1}{h_n} K\left(\frac{t-X_1}{h_n}\right) - E\left(\frac{1}{h_n} K\left(\frac{t-X_1}{h_n}\right)\right)\right|^{2q}\right) dt$$

$$= \varepsilon^{-2q} n^{-q} E \int_{-\infty}^{\infty} \left|\frac{1}{h_n} K\left(\frac{t-X_1}{h_n}\right) - E\left(\frac{1}{h_n} K\left(\frac{t-X_1}{h_n}\right)\right)\right|^{2q} dt$$

$$= \varepsilon^{-2q} n^{-q} E[2(\text{bdd}K)^{2q}(b - a)h_n(h_n)^{-2q}]$$

$$\leq C \frac{h_n}{(h_n^2 n)^q}.$$ (Here bddK is a bound on K and [a,b] is the compact support of K.)

Since $h_n = O(n^{-d})$, $0 < d < 1/2$, there exist q such that

$\sum_{n=1}^{\infty} h_n/(h_n^2 n)^q < \infty$ or $\sum_{n=1}^{\infty} P[||E(X_{n1}|U_{nn})|| > \varepsilon] < \infty$. ///

6.4 STRONG CONVERGENCE RESULTS FOR RANDOMLY WEIGHTED SUMS

The random nature of many problems in the applied sciences has caused many researchers to change from deterministic to probabilistic approaches. It then becomes important to consider random weighting of random variables. In this section a strong convergence result is obtained for randomly weighted sums of triangular arrays of random elements that are row-wise exchangeable in type p Banach spaces. Here $\{a_n\}$ and A_n are random weights and a_n is further assumed to be a symmetric function of $(X_{n1}, \ldots, X_{nn})$. Let U_{nn} and $U_{\infty n}$ be the σ-fields described in (6.3.9) and (6.3.10).

Theorem 6.4.1 Let $\{X_{nk}: 1 \leq k \leq n, n \geq 1\}$ be an array of row-wise exchangeable random elements in a type p Banach space, $1 \leq p \leq 2$. Let $\{a_n\}$ and $\{A_n\}$ be random weights where a_n is a symmetric function of $(X_{n1}, \ldots, X_{nn})$. Let U_{nn} and $U_{\infty n}$ be the σ-fields defined in (6.3.9) and (6.3.10). If

(i) $E||X_{nk} - X_{\infty k}||^2 = o(n^{-2\alpha}),\ \alpha = \frac{p-1}{p}$

(ii) $\rho_n(f) = E[f(X_{n1})f(X_{n2})] \to 0$ as $n \to \infty$ for $f \in E^*$,

(iii) $n^{1/p}|a_{n1}(\omega)| \leq M$ a.s., (6.4.1)

(iv) $A_n \geq M_1 > 0$ a.s. for $n \geq M_0$, (6.4.2)

(v) $||X_{n1} - X_{\infty 1}|| \geq ||X_{(n+1),1} - X_{\infty 1}||$ a.s. for all n,

then

$$||\frac{1}{A_n} \sum_{k=1}^{n} a_n X_{nk}|| \to 0 \text{ a.s.}$$

Proof: For $\varepsilon > 0$ and by (4.3.1)

$$P[\sup_{n \geq m} ||\frac{1}{A_n} \sum_{k=1}^{n} a_n X_{nk}|| > \varepsilon]$$

$$\leq P[\sup_{n \geq m} ||\frac{1}{A_n} \sum_{k=1}^{n} (a_n X_{nk} - a_n X_{\infty k})|| > \frac{\varepsilon}{2}]$$

$$+ P[\sup_{n \geq m} ||\frac{1}{A_n} \sum_{k=1}^{n} a_n X_{\infty k}|| > \frac{\varepsilon}{2}]$$

$$= P[\sup_{n \geq m} (\frac{n|a_n|}{A_n}) || (\frac{1}{n} \sum_{k=1}^{n} X_{nk} - \frac{1}{n} \sum_{k=1}^{n} X_{\infty k})|| > \frac{\varepsilon}{2}]$$

$$+ P[\sup_{n\geq m} (\frac{|a_n|}{A_n}) \,||\, \sum_{k=1}^{n} X_{\infty k} || > \frac{\varepsilon}{2}]$$

$$= P[\sup_{n\geq m} (\frac{n|a_n|}{A_n}) \,||\, E(X_{n1} - X_{\infty 1}|U_{nn})|| > \frac{\varepsilon}{2}]$$

$$+ P[\sup_{n\geq m} (\frac{|a_n|}{A_n}) \,||\, \sum_{k=1}^{n} X_{\infty k} || > \frac{\varepsilon}{2}]. \qquad (6.4.3)$$

From (6.4.1) and (6.4.2), for each n, $|a_n| \leq M_2 n^{-1/p}$ a.s. for some $M_2 > 0$ and $\frac{1}{A_n} \leq \frac{1}{M_1}$ a.s. Thus,

$$\frac{n|a_n|}{A_n} \leq M\, n^{\frac{(p-1)}{p}} \quad \text{a.s.} \qquad (6.4.4)$$

Substituting (6.4.4) into (6.4.3) gives

$$P[\sup_{n\geq m} (\frac{n|a_n|}{A_n}) \,||\, E(X_{n1} - X_{\infty 1}|U_{\infty n})|| > \frac{\varepsilon}{2}]$$

$$+ P[\sup_{n\geq m} (\frac{|a_n|}{A_n}) \,||\, \sum_{k=1}^{n} X_{\infty k} || > \frac{\varepsilon}{2}]$$

$$\leq P[\sup_{n\geq m} n^{(p-1)/p} || E(X_{n1} - X_{\infty 1}|U_{\infty n})|| > \frac{\varepsilon}{2M}]$$

$$+ P[\sup_{n\geq m} || n^{-1/p} \sum_{k=1}^{n} X_{\infty k} || > \frac{\varepsilon}{2M_2}].$$

From (i) of the hypothesis and by Lemma 6.3.3 it can be easily shown that $n^{\alpha}||E(X_{n1} - X_{\infty 1}|U_{\infty n})|| \to 0$ a.s.

$$P[\sup_{n\geq m} n^{\alpha} ||E(X_{n1} - X_{\infty 1}|U_{\infty n})|| > \frac{\varepsilon}{2M}] \to 0 \text{ as } n \to \infty.$$

By (6.1.1)

$$P[\sup_{n\geq m} ||n^{-1/p} \sum_{k=1}^{n} X_{\infty k}|| > \frac{\varepsilon}{2M_2}]$$

$$= \int_F P_\nu[\sup_{n\geq m} ||n^{-1/p} \sum_{k=1}^{n} X_{\infty k}|| > \frac{\varepsilon}{2M_2}] d\mu_\infty(P_\nu).$$

Since $\{X_{\infty k}\}$ are independent, identically distributed random elements with respect to (Ω, A, P_ν) and $E_\nu(X_{\infty 1}) = 0$, it follows by Theorem 6.3.8 that $||n^{-1/p} \sum_{k=1}^{n} X_{\infty k}|| \to 0$ a.s. Thus,

$P_\nu[\sup_{n\geq m} ||n^{-1/p} \sum_{k=1}^{n} X_{\infty k}|| > \frac{\varepsilon}{2M_2}] \to 0$ as $n \to \infty$ and

$\int_F P_\nu[\sup_{n\geq m} ||n^{-1/p} \sum_{k=1}^{n} X_{\infty k}|| > \frac{\varepsilon}{2M_2}] d\mu_\infty(P_\nu) \to 0$ as $n \to \infty$

by the bounded convergence theorem. Hence,

$$P[\sup_{n\geq m} ||\frac{1}{A_n} \sum_{k=1}^{n} a_n X_{nk}|| > \varepsilon] \to 0 \text{ as } n \to \infty$$

and

$$||\frac{1}{A_n} \sum_{k=1}^{n} a_n X_{nk}|| \to 0 \quad \text{a.s.}$$ ///

The next example is a version of Example 6.3.1, the general density estimation problem where $\{X_1, \ldots, X_n\}$ are independent, identically distributed random variables with the same density function f.

Example 6.4.1 Let $\{X_1, \ldots, X_n\}$ be independent, identically distributed random variables with the same density function f. The kernel estimate for f with variable bandwidth h_n is given by

$$f_n(t) = \frac{1}{n} \sum_{k=1}^{n} (1/h_n)K(\frac{t-X_k}{h_n})$$

where K is a bounded (integrable) kernel with compact support [a,b] and h_n is a bounded monotonically decreasing, symmetric function of $X_1, \ldots X_n$ such that for some c > 0 and $0 < \alpha < 1/2$, $(n + c)^{-\alpha} \leq h_n \leq n^{-\alpha}$ a.s.

As in Example 6.3.1, the condition $||E(X_{n1}|U_{\infty n})|| \to 0$ a.s. is crucial to the proof of Theorem 6.4.1. For $X_{nk} = \frac{1}{n} \sum_{k=1}^{n} \frac{1}{h_n} K(\frac{t-X_k}{h_n}) - E[\frac{1}{h_n} K(\frac{t-X_1}{h_n})]$, it is again easier to prove directly that $||E(X_{n1}|U_{\infty n})|| \to 0$ a.s.

6.5 PROBLEMS

6.1 In Theorem 6.1.1, prove (6.1.5).

6.2 In Theorem 6.3.1, prove that

$$\{x \in K: ||x|| > \eta\} = \bigcup_{i=1}^{t} \{x \in K: |f(x)| > \eta/2\}$$

where K is compact and $f_i \in E^*$, $1 \leq i \leq t$.

6.3 Prove Corollary 6.3.5 of Theorem 6.3.4.

6.4 Let $X_1, \ldots, X_n$ be random elements in a separable Banach space E. Show that if $P(||X_k|| > t) \leq P(||X_1|| > t)$ for all $k \geq 1$ and $t \in R$ then $E||X_k|| \leq E||X_1||$ for all $k \geq 1$.

BIBLIOGRAPHY

de Acosta, A.; Araujo, A.; Giné, E. (1978) Poisson measures, Gaussian measures and the central limit theorem in Banach spaces. Advances in Probability, Vol. IV, 1-68 (Kuelbs, Ed.) Dekker, New York.

Aldros, D. (1981). Representations for partially exchangeable arrays of random variables. J. Mult. Anal., 11, 581-598.

Aldous, D.J. (1977) Limit theorems for subsequences of arbitrarily-dependent sequences of random variables. Zeitschrift fur Wahrscheinlichkeitstheorie verw. Gebiete 40, 59-82.

Aldous, D.J. (1982) On exchangeability and conditional independence. In Exchangeability in Probability and Statistics (G. Koch and F. Spizzichino, eds.). North-Holland, Amsterdam, 165-170.

Aldous, D.J. and Eagleson, G.K. (1978) On mixing and stability of limit theorems. Ann. Probability 6, 325-331.

Anderson, T.W. and Taylor, J.B. (1976) Strong consistency of least square estimates in normal linear regression. Ann. Statist 4, 788-790.

Anis, A.A., and Gharib, M. On the variance of the maximum of partial sums of n-exchangeable random variables with applications. J. Appl. Probab. 17 (1980), 432-439.

Araujo, A. and Giné, E. (1980) The Central Limit Theorem for Real and Banach Valued Random Variables. Wiley, New York.

Bartfai, P. (1980) Remarks on the exchangeable random variables. Publ. Math. Debrecen 27, 143-148.

Beck, A. (1963) On the strong law of large numbers. Ergodic Theory. Academic Press, New York, 21-53.

Beck, A. (1976) Cancellation in Banach spaces. Proc. Probability in Banach spaces. LECTURE NOTES IN MATHEMATICS V526 Springer-Verlag, Berlin, 13-20.

Billingsley, P. (1968) Convergence of Probability Measures. Wiley, New York.

Blum, Chernoff, Rosenblatt, and Teicher (1958) Central limit theorems for exchangeable processes. Canadian J. of Mathematics, 10, 222-229.

Boes, D.C. and Salas, L.C.J.D. On the expected range and expected adjusted range of partial sums of exchangeable random variables. J. Appl. Probability 10 (1973), 671-677.

Bru, B., Heinich, H. and Lootgieter, J.C. (1981) Lois des grands nombres pour les variables exchangeables. C.R. Acad. Sci Paris. Math., 293, 485-488.

Bühlmann, Hans (1960) Austauschbare stochastische Variabeln und ihre Grenzwertsatze. Univ. of California Publications in Statistics, Vol. 3, No. 1.

Carnal, E. (1980) Independence conditionelle permutable. Ann. Inst. Henri Poincare B 16, 39-47.

Chanda, C. Asymptotic distribution of sample quantiles for exchangeable random variables. Calcutta Statist. Assoc. Bull. 20 (1971), 135-142.

Chernoff, H. and Teicher, H. (1958) A central limit theorem for sums of interchangeable random variables. Ann. Math. Stat. 29, 118-130.

Chow, Y.S. and Lai, T.L. (1973) Limiting behavior of weighted sums of independent random variables. Ann Prob., 1, 810-824.

Chow, Y.S. and Teicher, H. (1978) Probability Theory: Independence, Interchangeability, Martingales. Springer-Verlag, New York.

Chung, K.L. (1968) A Course in Probability. Harcourt, Brace and World, New York.

Cifarelli, D.M. and Regazzini, E. (1979) A general approach to Bayesian analysis of nonparametric problems. The associative mean values within the frame work of the Dirichlet process. Riv. Mat. Sci. Econom. Social., 2, 39-52.

Crabtree, J.C. (1975) Martingales in Banach spaces. Masters Thesis, University of South Carolina.

Dacunha, Castelle, D. A survey on exchangeable random variables in normed spaces. Exchangeability in Probability and Statistics (Rome, 1981), 47-60, North-Holland, Amsterdam-New York, 1982.

Daffer, P.Z. (1984) Central limit theorems for weighted sums of exchangeable random elements in Banach spaces. Stochastic Analysis & Appl's. 2, 229-244.

DeVroye, L.P. and Wagner, T.J. (1979). The L_1 convergence of kernel density estimates. Ann. Stat., 7, 1136-1139.

Diaconis, P. (1977) Finite forms of de Finetti's theorem on exchangeability. Synthese, 36, 271-281.

Diaconis, P. and Freedman, D. (1980) Finite exchangeable sequences. Annals of Probability, 8, 745-764.

Diaconis, P. and Freedman, D. (1980) de Finetti's theorem for Markov chains. Annals of Probability, 8, 115-130.

Diaconis, P. and Freedman, D. (1980) de Finetti's generalizations of exchangeability. Studies in Inductive Logic and Probability. Univ. of Calif. Press, 233-249.

Diestel, J. and Uhl. J.J. Jr. (1976) The Radon-Nikodym theorem for Banach space valued measures, Rocky Mountain J. Math. 6, 1-46.

Diestel, J. and Uhl, J.J. Jr. (1977) Vector Measures. Amer. Math. Soc., Math. Surveys Number 15.

Doob, J.L. (1940) Regular properties of certain families of chance variables. Trans. Amer. Math. Soc., 47, 455-486.

Dubins, L. and Freedman, D. (1964) Measurable sets of measures. Pacific Jour. of Math 14, 1211-1222.

Dubins, L.E. and Freedman, D. (1979) Exchangeable processes need not be mixtures of independent, identically distributed random variables. Z. Wahr. verw. Geb., 48, 115-132.

Dubins, L.E. (1983) Some exchangeable probabilities which are singular with respect to all presentable probabilities. Zeit. fur Wahr. verw. Geb., 64, 1-6.

Dunford, N. and Schwartz, J.T. (1963) Linear Operators. I. New York: Interscience.

Eagleson, G.K. and Weber, N.C. (1978) Limit theorems for weakly exchangeable arrays. Math. Proc. Camb. Phil Soc., 84, 123-130.

Eagleson, G.K. (1982) Weak limit theorems for exchangeable random variables. Exchangeability in Probability and Statistics (Rome, 1981), 251-268. North-Holland, Amsterdam-New York, 1982.

Edgar, G.A. and Sucheston, L. (1981) Demonstrations de lois des grands nombres par les sous-martingales descendantes. C. R. Acad. Sci. Paris Math. 292, 967-969.

Feller, W. (1971) An Introduction to Probability Theory and Its Application, Vol. II Wiley Publishers, New York, 228-230, 287, 423.

de Finetti, B. (1972) Probability, Induction and Statistics. Wiley, New York.

de Finetti, B. (1974) Theory of Probability. Wiley, New York.

Freedman, D.A. (1963) Invariants under mixing which generalize de Finetti's Theorem. Ann. Math. Stat. 34, 916-923.

Freedman, D.A. (1980) A mixture of independent identically distributed random variables need not admit a regular conditional probability given the exchangeable σ-field. Z. Wahr. verw. Geb., 51, 239-248.

Freedman, D. and Diaconis, P. (1982) de Finetti's theorem for symmetric location families. Annals. of Statistics, 80, 184-189.

Galambos, J. The role of exchangeability in the theory of order statistics. Exchangeability in Probability and

and Statistics (Rome, 1981), 75-86. North-Holland, Amsterdam, New York, 1982.

Grenander, U. (1963) Probabilities on Algebraic Structures. Wiley, New York.

Hall, P. and Heyde, C.C. (1980) Martingale Limit Theory and Its Application. Academic Press.

Halmos, P.R. (1950) Measure Theory. Van Nostrand, Princeton.

Heath and Sudderth (1976) de Finetti's theorem on exchangeable random variables. Amer. Statistician, 30, 188-189.

Hewitt, E. and Savage, L. (1955) Symmetric measures on cartesian products. Trans. Amer. Math. Soc. 80, 470-501.

Hille, E. and Phillips, R. (1957). Functional Analysis and Semi-groups. Amer. Math. Soc. Colloq. Publ. Vol. XXXI. Providence, Rhode Island.

Hoffmann-Jørgensen, J. (1970) The Theory of Analytic Spaces. Various Publications Series No. 10. Matematisk Institut, Aarhus Universitet.

Hoffmann-Jørgensen, J. and Pisier, G. (1976) The law of large number and the central limit theorem in Banach spaces. Annals of Probability, 4, 587-599.

Hoover, D.N. Row-column exchangeability and a generalized model for probability. Exchangeability in Probability and Statistics (Rome, 1981), 281-291. North-Holland, Amsterdam-New York, 1982.

Hu, Y.S. (1979) A note on exchangeable events. J. Appl. Prob., 16, 662-664.

Johnson, W.B., Schechtman, G. and Zinn, J. (1985). Best constants in moment inequalities for linear combinations of independent and exchangeable random variables. Annals, Probability, 13, 234-253.

Kallenberg, O. (1973) Canomical representations and convergence criteria for processes with interchangeable increments. Zeit. Wahr. verw. Geb. 27, 23-36.

Kallenberg, O. (1974) Path properties of processes with independent and interchangeable increments. Zeit. fur Wahr. verw Geb., 28, 257-271.

Kallenberg, O. (1982) A dynamical approach to exchangeability. In: Exchangeability in Probability & Statistics, 87-96, Koch & Spizzichino, Ed.'s. North Holland.

Kendall, D.G. (1967) On finite and infinite sequences of exchangeable events. Stud. Scient. Math. Hungarria 2, 319-327.

Kingman, J.F.C. (1978) Uses of exchangeability. Ann. Prob., 6, 183-197.

Kingman, J.F.C. (1980) Mathematics of genetic diversity. Society for Industrial and Applied Mathematics (SIAM), Philadelphia, Pa..

Kingman, J.F.C. (1982) Exchangeability and the evolution of large populations, Exchangeability in Probability and Statistics (Rome, 1981) 97-112. North-Holland, Arsterdam-New York, 1982.

Leon, R.V. and Proschan, F. (1979) An inequality for convex functions involving G-majorization. J. Math. Anal. Appl. 69, 603-606.

Loeve, M. (1963) Probability Theory. Van Nostrand.

Longnecker, M. and Serfling, R.J. (1978) Moment inequalities for S$sub in under general dependence restrictions, with applications. Z. Wahrsch. Verw. Gebiete 43, 1-21.

Magiera, R. (1980) Estimation problem for the exponential class of distributions from delayed observations. Internat. Conf., Wisla, 1978, 275-287, Lecture Notes in Statist., 2. Springer, New York.

Mandrekar, V. and Zinn, J. (1980) Central limit problem for symmetric case: Convergence to non-Gaussian laws. Studia Mathematica, T. LXVII, 279-296.

Moran, P.A.P. (1973) A central limit theorem for exchangeable variates with geometrical applications. J. Appl. Prob., 10, 837-846.

Moscovici, E. and Popescu, O. (1973) A theorem on exchangeable random variables. Stud. Cerc. Mat. 25, 379-383.

Mourier, E. (1953) Elements aleatories dan un espace de Banach. Ann. Inst. Henri Poincare', 13, 159-244.

Olshen, R.A. (1971) The coincidence of measure algebras under an exchangeable probability. Zeit. Wahr. verw. Geb., 18, 153-158.

Olshen, R, (1973-1974) A note on exchangeable sequences. Zeit. Wahr. Verw. Geb., 28, 317-321.

Parzen, E. (1962) On estimation of a probability density and mode. Ann. Math. Statist., 33, 1065-1076

Pruitt, W. (1966) Summability of independent random variables. J. Math & Mech. 15, 769-776

Rao, C.R. and Srivastava, R.C. (1979) Some characterizations based on a multivariate splitting model. Sankhya Ser. A. 41, 124-128

Rohatgi, V.K. (1976) An Introduction to Probability Theory and Mathematical Statistics. Wiley, New York.

Rohatgi, V.K. (1968) Convergence rates in the law of large numbers II. Proc. Cambridge Philos, Soc. 64, 485-488.

Rosenblatt, M. (1971) Curve estimates. Ann. Math. Statist., 42, 1815-1842.

Royden, H.L. (1972) Real Analysis, 2nd Ed., MacMillan, New York.

Saunders, R. (1976) On joint exchangeability and conservative processes with stochastic rates. J. Appl. Prob., 13, 584-590

Scalora, F.S. (1961) Abstract martingale convergence theorems. Pacific J. Math., 11, 347-374.

Shaked, M. (1977) A concept of positive dependence for exchangeable random variables. Ann. Statist. 5, 505-515.

Szczotka, W. A characterization of the distribution of an exchangeable random vector. Bull. Acad. Polon, Sci. Ser. Sci. Math. 28 (1980), 411-414 (1981).

Taylor, R.L. (1979) Stochastic Convergence of Weighted Sums of Random Elements in Linear Spaces. Lecture Notes in Mathematics Vol. 672. Springer-Verlag.

Taylor, R.L. and Wei, D. (1978) Convergence of weighted sums of tight random elements. J. Mult. Anal., 8, 282-294.

Taylor, R.L. and Wei, D. (1979) Laws of large numbers for tight random elements in normed linear spaces. Ann. Probability, 7, 150-155.

Taylor, R.L. (1982a) Weak laws of large numbers for exchangeable random variables. USC Statistics Technical Report No. 88, University of South Carolina.

Taylor, R.L. (1982b) Convergence of weighted sums of arrays of random elements in type p spaces with application to density estimation. Sankhya, A, 44, 341-351.

Taylor, R.L. (1983) Complete convergence of weighted sums of arrays of random elements. Int. J. Math. & Math. Sci., 6, 69-79.

Taylor, R.L. (to appear) Laws of large numbers for dependent random variables. Colloquia Mathematica Societatis János Bolyai 36. Limit Theorems in Probability and Statistics, Veszprem (Hungary), 1982. 1023-1042

Teicher, H. (1971) On interchangeable random variables. Studi di Proabilita Statistica e Ricerca in Onore di Giuseppe Pompilj, 141-148.

Teicher, H. Renewal theory for interchangeable random variables. Exchangeability in Probability and Statistics (Rome, 1981), 113-121. North-Holland, Amsterdam-New York, 1982.

Troutman, B.M. (1983) Weak convergence of the adjusted range of cumulative sums of exchangeable random variables. J. Appl. Prob., 20, 297-304

Tucker, H.G. (1967) A Graduate Course in Probability. Academic Press, New York.

Vakhania, N.N. (1981) Probability Distribution on Linear Spaces. Elsevier-North Holland.

Vershik, A.M. and Schmidt, A.A. (1977) Limit measures arising in the theory of symmetric groups I. Theory Prob. Appl., 22, 70-85

Weber, N.C. (1980) A martingale approach to central limit theorems for exchangeable random variables. Jour. of Applied Prob. 17, 662-673.

Wei, D. and Taylor, R.L. (1978) Geometric consideration of weighted sums, convergence and random weighting. Bull. Inst. Math. Acad. Sinica. 6, 49-59.

Wilansky, A. (1964) Functional Analysis. Blaisdell, New York.

Woyczynski, W. (1978) Geometry and martingales in Banach spaces, Part II. Advances in Probability, 4, Dekker, 267-517.

Woyczynski, W. (1980) On Marcinkiewicz-Zygmund laws of large numbers in Banach spaces and related rates of convergence. Probab. and Math. Statist., 1, 117-131.